MODELLING HUMAN-FLOOD INTERACTIONS

A COUPLED FLOOD-AGENT-INSTITUTION MODELLING FRAMEWORK FOR LONG-TERM FLOOD RISK MANAGEMENT

Yared Abayneh Abebe

Cover design	Juliana Duque (juliana.es.duque@gmail.com)
Layout	Yared Abayneh Abebe, LaTeX

MODELLING HUMAN-FLOOD INTERACTIONS

A COUPLED FLOOD-AGENT-INSTITUTION MODELLING FRAMEWORK FOR LONG-TERM FLOOD RISK MANAGEMENT

DISSERTATION

Submitted in fulfilment of the requirements of
the Board for Doctorates of Delft University of Technology
and
of the Academic Board of the IHE Delft
Institute for Water Education
for
the Degree of DOCTOR
to be defended in public on
Thursday, 3 December 2020 at 12:30 hours
in Delft, the Netherlands

by

Yared Abayneh ABEBE

Master of Science in Water Science and Engineering; Specialization
Hydroinformatics,
IHE Delft Institute for Water Education

born in Addis Ababa, Ethiopia

This dissertation has been approved by the
promotor: Prof. dr. D. Brdjanovic and
copromotor: Dr. Z. Vojinovic

Composition of the doctoral committee:

Rector Magnificus TU Delft	Chairman
Rector IHE Delft	Vice-Chairman
Prof.dr. D. Brdjanovic	TU Delft/IHE Delft, promotor
Dr. Z. Vojinovic	IHE Delft, copromotor

Independent members:

Prof.dr. R.W. Künneke	TU Delft
Prof.dr.ir. A.E. Mynett	TU Delft/IHE Delft
Prof.dr. J.P. O'Kane	University College Cork, Ireland
Dr. O. Mark	Kruger A/S, Denmark
Prof.dr.ir. N.C. van de Giesen	TU Delft, reserve member

This research was conducted under the auspices of the Graduate School for Socio-Economic and Natural Sciences of the Environment (SENSE)

CRC Press/Balkema is an imprint of the Taylor & Francis Group, an informa business

Published by:
CRC Press/Balkema
Schipholweg 107C, 2316 XC, Leiden, the Netherlands
Pub.NL@taylorandfrancis.com
www.crcpress.com — www.taylorandfrancis.com
ISBN 978-0-367-74886-9

Contents

LIST OF FIGURES

List of Tables

ACKNOWLEDGEMENTS

Finally, the challenging journey of my PhD research comes to an end. A number of people have played an invaluable role in the process that leads to this dissertation, providing much needed academic, financial and emotional support.

I would like to acknowledge my supervisors, who coached me to become a matured, independent researcher. My promotor, Damir Brdjanovic, I thank you for all the support you provided me given the circumstances. I am very grateful for the expertise, encouragement and continuous financial support shown by the copromotor Zoran Vojinovic. You initiated the research idea, secured the fund and trusted me to carry out the research. You always have positive, praising words towards my work, and I very much appreciate that. I cherish our discussions on subjects ranging from refining the research aim and methodology to tennis matches and music.

My supervisors Igor Nikolic and Amineh Ghorbani played a substantial role in this dissertation and my scientific development. You introduced me to the world of agent-based modelling and put all your effort to help me master this paradigm of modelling. Igor, talking to you always fills me with energy and confidence. Your passion for research is contagious. Amineh, you always had time for my meeting requests to discuss all kinds of modelling and publication issues. I value your dedication and guidance. I thank you both for the advice and support throughout the research period.

I would like to thank Arlex Sanchez for the valuable discussions that shaped the research aim, especially at the beginning of the research. I am grateful to the independent doctoral committee members who took time to read and assess the dissertation, and are willing to be part of the defence ceremony.

The PhD research was made possible by the financial support from the research projects PEARL (Preparing for Extreme And Rare events in coastal regions), funded by the European Union Seventh Framework Programme (FP7) under Grant agreement n° 603663. PEARL also financially supported the fact-finding and needs assessment mission that I participated in after Hurricane Irma ruined Sint Maarten in 2017. I took part in the mission during my final year of the regular PhD time frame. Although it delayed my PhD graduation, witnessing the destruction the hurricane brought boosted my endeavour in shaping societies resilience to hydrometeorological hazards, as well as improving my research directly. To that end, I express my gratitude to our Sint Maarten partners, especially Paul Marten, for providing data for the case study.

Working within the PEARL consortium gave me a chance to collaborate with several researchers. I would like to thank all project members, especially Arabella Fraser, Natasa Manojlovic and Angelika Grhun for providing data, and the PEARL young researchers for the company and laughter during project meetings. The PhD research also received financial support from the RECONECT (Regenerating ECOsystems with Nature-based solutions for hydro-meteorological risk rEduCTion), from the European Union's Horizon 2020 Research and Innovation Programme under Grant Agreement n° 776866.

I thank SURFsara, especially Ander Astudillo, for providing a High Performance Computing (HPC) cloud resources that we used to run all the simulations. I also thank the Repast Simphony development team at Argonne National Laboratory, especially Eric Tatara and Nick Collier, for their support on the agent-based model development. I sincerely acknowledge those who ask and, most importantly, answer questions on online platforms. Their posts saved valuable time and prevented frustrations.

An emotional roller coaster characterizes the PhD research life. In that regard, the support of family and friends is indispensable. Neiler and Alida, apart from the friendship, working with you in the PEARL project was exciting. Thank you for the camaraderie and support. Neiler, I had enjoyed your company in project meetings and missions, and I enjoy writing articles with you. My dear friends — Pato, Aki, Angie, Can, Jessi, Mauri, Juan Carlos, Kun, Vero, Nata, Miguel, María Fernanda, Juan Pablo, Alex, Benno, Maribel, Stefan, Mohanned, Pablo, Mario, Adele, Irene, Thaine, Ana María — it is a pleasure having you in my life. We have spent several social gatherings, and academic discussions with some of you, that truly improve the quality of my life. I also thank my friends in Ethiopia: Reze, Ashu, Mafi, Mike, Dave, Bini and Mati for their friendship and encouragement. I would like to appreciate the "Habesha buddies" for their friendship.

I acknowledge the support from current and graduated IHE PhD fellows and Ethiopian PhD and MSc students at IHE. I thank Jolanda Boots and Anique Karsten for their help regarding administrative issues. I also thank all participants of the Saturday football practices for the fun we had. I am very grateful for the Water Engineering and Management group at the University of Twente, especially Suzanne Hulscher for permitting and Filipe, Koen, Joost and Sara for sharing their offices in the second half of 2019 and early 2020. I commend the group for its friendly and welcoming atmosphere.

At last, I would like to thank my family. Juliette, you are always there for me. You have been the academic advisor at home. We have brainstormed and discussed in detail to tackle research-related issues, which range from how to better illustrate simulation results to methodological hurdles and scientific publication dilemmas. I thank you for your love and continuous support. About a year ago, a new member joined the family, our beloved Eliana, whose presence improved our life significantly. My family in Ethiopia are my inspiration. My parents (Kasanesh Mengiste and Abayneh Abebe) and siblings (Kido, Hani, Dave and Samuel), I thank you for the love, encouragement and prayers. Your presence is always comforting. I am grateful to my in-laws, especially Yolanda Arevalo, who has been very helpful.

SUMMARY

Flooding disasters are the most common weather-related disasters affecting millions of people and causing economic damages in billions of dollars each year. With an increase in flood events, urban areas are particularly more affected. The negative impacts of floods are attributed to the extent and magnitude of a flood hazard, and the vulnerability and exposure of natural and human elements. In flood risk management (FRM) studies, flood modelling using hydrodynamic models has been the mainstream practice. However, these models analyse only one side of the coin, which is investigating the frequency and magnitude (i.e., depth, velocity and extent) of flooding. As a result, risk reduction strategies focus on engineering structural measures and hazard-based risk awareness and warning systems. These models disregard the effects of economic, social, cultural, institutional and governance factors on flood hazard, vulnerability and exposure.

Recently, a socio-hydrologic approach that integrates all components of risk is being promoted to strengthen FRM and to reduce flood risk. This approach should consider the interactions between human and flood subsystems across multiple spatial, temporal and organizational scales. To that end, researchers have formalized and modelled subsystems' processes using differential equations. Although these methods are easy to use and flexible, they do not address the heterogeneity that exists within the human subsystem, and they do not incorporate the institutions that shape the behaviour of individuals.

Addressing these gaps, the main objectives of the dissertation are to develop a modelling framework and a methodology to build models for holistic FRM, and to assess how coupled human-flood interaction models support FRM policy analysis and decision-making. To achieve the objectives, we first develop a modelling framework called Coupled fLood-Agent-Institution Modelling framework (CLAIM). CLAIM integrates actors, institutions, the urban environment, hydrologic and hydrodynamic processes and external factors which affect local FRM activities. The framework draws on the complex system perspective and conceptualizes the interactions among floods, humans and their environment as drivers of flood hazard, vulnerability and exposure.

In the methodology that accompanies the CLAIM framework, the human subsystem is modelled using agent-based models (ABMs). Consequently, CLAIM incorporates heterogeneous actors and their actions and interactions with the environment and flooding. It also provides the possibility to analyse the underlying institutions that govern the actions and interactions in managing flood risk. The flood subsystem is modelled using a physically-based, numerical model. The ABM is dynamically coupled to the flood model to understand human-flood interactions and to investigate the effect of institutions on FRM policy analysis.

Applications of the modelling framework were explored in actual case studies.

The first case is the Caribbean island of Sint Maarten, which is selected to explore the implications of mainly formal FRM institutions. The second one is the case of Wilhelmsburg, a quarter in Hamburg, Germany, which is used to explore the effects of informal institutions on household adaptation measures to reduce vulnerability to flooding.

This dissertation provides three main scientific contributions. It contributes to socio-hydrology by providing a framework that allows a holistic conceptualization and modelling of the human-flood interactions. The framework defines system elements that should be considered during conceptualization and explicitly incorporates institutions that drives flood risk. Further, the research contributes to social and hydrological knowledge integration which facilitates interdisciplinary research. Finally, it contributes to FRM by providing a holistic view of flood risk in which one could study how the social, economic, governance and hydrological makeup of an area affect the risk. The coupled ABM-flood models allow to study how levels of flood hazard, exposure and vulnerability change simultaneously with changes in human behaviour. The coupled models also provide a platform to test existing and proposed policies for flood risk reduction .

SAMENVATTING

Overstromingen behoren tot de meest voorkomende weer-gerelateerde rampen die jaarlijks miljoenen mensen treffen en miljarden aan kosten met zich meebrengen. Door de toename van overstromingen zullen met name stedelijke gebieden meer worden getroffen. De negatieve gevolgen van overstromingen zijn afhankelijk van de omvang en grootte van een overstromingsramp en de kwetsbaarheid van natuurlijke en menselijke elementen. Bij studies naar het beheersen van overstromingsrisico's worden tegenwoordig veelal hydrodynamische numerieke modellen gebruikt. Echter, deze modellen belichten slechts een kant van de medaille, namelijk inzicht in frequentie en grootte (d.w.z. waterdiepte, snelheid en omvang) van overstromingen. Dientengevolge richten strategieën voor overstromingsbeperking zich vaak op structurele maatregelen en waarschuwings-systemen voor calamiteiten. Deze modellen houden geen rekening met andere effecten zoals economische schade, sociale-culturele- of institutionele overwegingen, en bestuurlijke aspecten die zich voordoen bij blootstelling aan overstromingen.

Een socio-hydrologische aanpak die al deze risico aspecten integreert vindt recent weerklank bij het beheersen van overstromingsrisico's en het verminderen van de gevolgen. Een dergelijke benadering gaat uit van de interacties tussen mensen en overstromingen op meerdere ruimtelijke, temporele en organisatorische schalen. Daartoe hebben onderzoekers het menselijk gedrag in verschillende subsystemen geformaliseerd en uitgedrukt in wiskundige differentiaalvergelijkingen. Hoewel gemakkelijk en flexibel in gebruik, adresseren deze methoden niet de heterogeniteit van menselijke subsystemen en bevatten zij niet de institutionele aspecten die het gedrag van mensen beïnvloeden.

De belangrijkste doelstelling van dit proefschrift is om deze beperkingen op te heffen en een modeleerraamwerk en methodiek te ontwikkelen voor een holistische aanpak tot het beheersen van overstromingsrisico's, en na te gaan hoe gekoppelde mens-overstroming interactiemodellen kunnen bijdragen aan beleid en besluitvorming in geval van overstromingsrisico's. Om dit doel te bereiken is eerst een raamwerk ontwikkeld genaamd Coupled fLood-Agent-Institution Modelling framework (CLAIM). CLAIM integreert actors, instituties, de stedelijke omgeving, hydrologische en hydrodynamische processen en externe factoren die de lokale overstromingsrisicomodellen beïnvloeden. Het raamwerk gaat uit van een benadering op basis van complexe systeemtheorie met oog voor de interacties tussen overstromingen, mensen, en hun omgeving in geval van overstromingsgevaar.

In de methodiek van het CLAIM raamwerk wordt het menselijk subsysteem gemodelleerd met behulp van agent-based models (ABMs). Het gevolg hiervan is dat CLAIM de mogelijkheid biedt om de heterogeniteit van een populatie en hun acties te representeren inclusief de interacties met de overstroming in hun omgeving. Het biedt tevens de mogelijkheid om de onderliggende beweegredenen te analyseren die hun acties en interacties in geval van overstromingsrisico's bepalen. Het overstromingssubsysteem wordt gemodelleerd met behulp van een numeriek modelsysteem dat dynamisch wordt gekoppeld met het menselijk subsysteem teneinde de interacties tussen overstromingen en menselijk gedrag te begrijpen en het effect van instellingen op beleidsanalyse te onderzoeken.

Het raamwerk is getoetst aan de hand van feitelijke situaties. De eerste toepassing betreft het Caribbische eiland Sint Maarten waar de implicaties van het hebben van formele instituties op het gebied van overstromingsrisico worden nagegaan. De tweede toepassing betreft Wilhelmsburg, een wijk in Hamburg, Duitsland, waar informele instituties en adaptatiemaatregelen worden gebruikt om de kwetsbaarheid tegenover overstromingen te beperken.

Dit proefschrift bevat drie belangrijke wetenschappelijke bijdragen. Het versterkt het vakgebied van de socio-hydrologie door een raamwerk te verschaffen dat een holistische benadering op basis van gekoppelde modellen toestaat voor de interacties tussen overstromingen en menselijk handelen. Het raamwerk bestaat uit systeem elementen die van belang zijn voor de modelvorming en houdt expliciet rekening met instellingen die zich richten op overstromingsrisico's. Daarnaast draagt dit onderzoek bij aan de integratie van sociale en hydrologische kennis en faciliteert het interdisciplinair onderzoek. Tot slot draagt het bij aan het beheersen van overstromingsrisico's doordat het een holistische aanpak mogelijk maakt waarin onderzocht kan worden hoe de sociale, economische, bestuurlijke, en hydrologisch componenten daaraan bijdragen. Het gekoppelde modelsysteem maakt het mogelijk om na te gaan hoe verschillende niveaus van overstromingsgevaar en kwetsbaarheid veranderen in samenhang met veranderingen in menselijk gedrag. Het ontwikkelde CLAIM raamwerk verschaft een platform waarmee bestaande maatregelen kunnen worden getest en nieuwe richtlijnen kunnen worden opgesteld om overstromingsrisico's te beperken.

List of abbreviations

1D	One dimensional
2D	Two dimensional
ABM	Agent-based model or modelling
ADICO	Attribute, deontic, aim, condition, or else
BO	Building and housing ordinance
BP	Beach policy
CAS	Complex adaptive system
CLAIM	Coupled flood-agent-institution modelling framework
CN	Curve number
FRM	Flood risk management
FZ	Flood zoning policy
GIS	Geographic information system
MAIA	Modelling agent systems using institutional analysis
NDP	National development plan
ODD	Overview, design concepts, and details
OFAT	One-factor-at-a-time
PEARL	Preparing for extreme and rare events in coastal regions
PMT	Protection motivation theory
SA	Sensitivity analysis
UV	Uncertainty analysis
VROMI	a Dutch acronym for Ministry of Public Housing, Spatial Planning, Environment and Infrastructure

1

Introduction

1.1 Motivation

Of all weather-related disasters in the last two decades, floods are by far the most common (47 %), affecting 2.3 billion people (CRED and UNISDR, 2015). The CRED and UNISDR report emphasizes that after storms and geophysical disasters, floods have been causing the third highest amount of economic damage (662 billion USD) over the past 20 years. The number of flood events has significantly increased, in which urban areas have been hit particularly hard (Jha *et al.*, 2012). The risk associated with floods can be defined as the probability of negative impacts due to floods (Schanze, 2006). Flood impacts are mainly attributed to the extent and magnitude of a flood hazard which can be caused by one or a combination of fluvial, flash, pluvial, groundwater and coastal floods (Vojinovic and Huang, 2014). However, the negative impacts are also due to the vulnerability and exposure of natural and human elements such as individuals, livelihoods, economic and cultural assets, infrastructure, ecosystems and environmental resources (Vojinovic *et al.*, 2016).

In his dissertation, Gilbert F. White (1945, p. 2) states: "Floods are 'acts of God,' but flood losses are largely acts of man." One may argue that floods can be "acts of human" as much as they are "acts of God." For example, a rainfall with certain intensity may cause flooding that disturbs livelihoods in an Ethiopian city due to poor drainage infrastructure while a Dutch city may not register flooding from an equivalent rain intensity. Nevertheless, White's statement that flood losses are aggravated by human encroachment of floodplains is indisputable.

Furthermore, in an article entitled "Taking the naturalness out of natural disasters," O'Keefe *et al.* (1976, p. 566) stated: "Without people there is no disaster," asserting their argument that socio-economic factors contribute to disasters more than natural factors do. When "nature" is considered as the threat, the hazard component of a disaster becomes more influential, and risk reduction strategies focus on engineering structural measures and hazard-based risk awareness and warning systems (Gaillard, 2010).

But, such measures are not always effective. For example, structural measures such as dykes are designed based on return periods (e.g., a 100-year storm event), which are computed using statistical analysis of historical flood data in an area. This entails that dykes may fail or overtopped when a potentially higher flood event occurs or when peak flood of the design return period increases over time due to, for example, climate change (Pinter *et al.*, 2016). Such scenarios are sources of residual risk that residents either unaware of or ignore. As a result, dykes creates a sense of safety by reducing residents' flood risk perception (Ludy and Kondolf, 2012). Ludy and Kondolf conclude that residents become "involuntarily exposed to risk".

Based on these arguments, comprehensive approaches forwarded to reduce flood impacts should include human adjustment to floods (White, 1945), and focus on human elements such as vulnerability, capacity and resilience, which are shaped by socio-economic factors (Gaillard, 2010; O'Keefe *et al.*, 1976).

1.2 HUMAN-FLOOD INTERACTIONS

Floods and their impacts are not just nature-related. They rather are the result of meteorological and hydrological factors aggravated by human actions (APFM, 2012). Changes in the climate system and economic, social, cultural, institutional and governance factors are drivers of flood hazard, vulnerability and exposure (IPCC, 2012, 2014a). For example, in the context of urban flood risk, population growth and the associated urban expansion result in changes to land use and land cover. That leads to an increase in impermeable surfaces, which increases the flood hazard. When accompanied by inadequate planning and policies, urban expansion may occur in flood-prone areas increasing exposure; or it may occur in dense, low-quality informal settlements that contribute to a higher number of vulnerable people (Jha *et al.*, 2012). For example, in the UK, the government acknowledged that the increasing demand for housing leads to more building in high flood risk zones (Department for Communities and Local Government, 2007), in which the proportion of new residential properties located in flood zones grow from 7 % in 2013-14 to 9 % in 2015-16 (Department for Communities and Local Government, 2016).

Moreover, the behaviour of individuals plays an essential role in flood risk. Based on their economic situation and risk perception[1], heterogeneous individuals living in flood-prone areas may implement their local measures to reduce hazard (e.g. nature based solutions such as green roofs or rainwater tanks (Vojinovic and Huang, 2014)) or vulnerability (e.g. dwellings with a non-habitable ground floor (Gersonius *et al.*, 2008)). Further, individuals may insure their properties to avoid financial losses or to recover better, in the case of flooding. Currently, governments are reorganizing flood insurance policies changing individual behaviour (Dubbelboer *et al.*, 2017). Individuals may also reduce exposure to flood hazard by relocating assets to less flood-prone areas and through evacuations (UNISDR, 2015).

In flood risk management (FRM), the likelihoods of adopting and implementing measures that reduce flood hazard, vulnerability and exposure depend on changes in individual and institutional behaviour in response to the potential of flooding

[1] Risk perception is a function of values, feelings, experiences and cultural perspectives (Schanze, 2006).

and the accompanying impact (Loucks, 2015). Therefore, on the one hand, FRM is dependent on the rules, regulations, policies and implementations that aim to reduce flood risk. On the other hand, it relies on how individuals react towards those aspects and adapt their behaviour. The factors, which shape the flood hazard and a community's exposure and vulnerability to flooding, can be understood as *institutions*. Institutions are key elements in the social, economic and political makeup of human beings that define our interactions with the physical system. The importance of institutions as social structures that influence society as a whole, and in turn, are influenced by society has been repeatedly emphasized by prominent scholars in economics, political science, sociology and ecology among others (e.g., Hodgson, 1988; North, 1990; Ostrom, 1990; Young, 1986).

1.3 Systems perspective and sociohydrology

To strengthen FRM and to reduce flood risk, a holistic, interdisciplinary approach that integrates all components of risk is essential (Aerts *et al.*, 2018). This approach should consider the interactions between human and physical subsystems[2] (Schanze, 2006; Vojinovic, 2015). The "human subsystem" consists of decision-making individuals, whose collective behaviour creates and is constrained by institutions such as norms, habits and laws. The human subsystem is embedded in and interacts with the "physical subsystem". The physical subsystem includes drainage systems and dykes that might be affected by flood events, and the flood itself. With interactions across multiple spatial, temporal and organizational scales, and behaviour driven by imperfect information and bounded rationality, the coupled human-flood system is a *complex system* (see also Pahl-Wostl, 2015). Further, as individuals and organizations learn (Mitchell, 2009) from previous flood impacts, the human-flood system is a *complex adaptive system (CAS)*.

Human-flood interaction studies have been a subject of interest for decades. However, there has been resistance from hydrologists to include or couple models that capture the human dynamics with their hydrological models (Loucks, 2015). As a result, models used for policy decision support in FRM focuses on quantitative assessment of flood hazard and flood hazard reduction. Recently, modelling of the coupled human-flood system is getting more attention in *socio-hydrology* (also *sociohydrology*), which studies the co-evolution of humans and water explicitly by considering the possibility of generating emergent behaviours (Sivapalan *et al.*, 2012). In socio-hydrology, the human subsystem is regarded as an endogenous part of the water subsystem, and there is a two-way interaction between the two subsystems.

[2]The term "physical" in the physical subsystem is a generic expression. Depending on the coupled model we address, as in Chapter 2, it will be replaced by a specific term (for example, "environment" or "ecology" in socio-environmental or socio-ecological systems; "technology" in socio-technical systems; "water" in coupled human-water systems; and "flood" in coupled human-flood systems).

1.4 Research gaps in human-flood interaction modelling

Sivapalan and Blöschl (2015) identified two possible approaches to model coupled human-flood interactions. The first ones are called *stylized models*[3], and they formalize the human and flood subsystems processes using a single differential equation. The second type of models are called *comprehensive system-of-systems models*, and they represent the subsystems by individual models that are based on well-established methodologies from the relevant disciplines.

Stylized models such as those developed by Ciullo *et al.* (2017), Di Baldassarre *et al.* (2013, 2015) and Viglione *et al.* (2014) conceptualize the dynamics of settled floodplains as a complex human-flood system. These models are easy to use and flexible. But, as also pointed out by the above authors, the main drawback of the stylized models is that they neglect the heterogeneity that exists within the human subsystem. In addition, their conceptualization is based only on societal memory or experience of prior flood events as a link between the human and flood subsystems. The model conceptualization does not incorporate the institutions that shape the behaviour of humans in their interactions with their environment and flood.

The only stylized human-flood model that considered institutions is the study by Yu *et al.* (2017). Yu *et al.* studied human-flood interactions in polders of coastal Bangladesh by including institutions for collective actions. But, they also used stylized models that do not consider heterogeneity, and focused only on informal institutions for collective actions.

Conversely, studies such as (Dawson *et al.*, 2011; Dubbelboer *et al.*, 2017; Erdlenbruch and Bonté, 2018; Haer *et al.*, 2016; Tonn and Guikema, 2017) developed system-oriented models that conceptualize and model human-flood interactions using agent-based models (ABMs) considering decision makings of heterogeneous actors. However, one of the main gaps in these studies is that they either consider flood as an exogenous element (for example, an agent's flood experience is set initially and stays the same throughout the simulations) or simplify flood models. Another gap is that they do not methodically analyse institutions to study drivers of flood risk. Instead, they use simplified set of behavioural rules.

Votsis (2017) utilized a cellular automaton model to study the relationship between urbanization trends and FRM strategies. The study shows the effects of bottom-up, flood risk information-based housing market responses and top-down floodplain development restriction scenarios on urbanization. However, the study does not show if the flood extent and depth changes with the development pattern. It also focuses only on the exposure component of the flood risk.

In general, there are important initiatives to model human-flood interactions using a systems perspective. However, these efforts are fragmented and do not address either heterogeneity of actors or all components of the flood risk (i.e., hazard,

[3]Stylized models are also referred to as "system dynamic models" (Konar *et al.*, 2019) and "conceptual models" (Troy *et al.*, 2015b).

vulnerability and exposure) in their modelling exercise. Besides, a systems approach which explicitly takes into account institutions as factors that shape the flood hazard and community's exposure and vulnerability to flooding has not yet been sufficiently addressed in the literature. Developing a framework that integrates the human and flood subsystems and supports modelling decision makings of multiple stakeholders in FRM has also been a major challenge (O'Connell and O'Donnell, 2014).

1.5 Research aim and questions

The aim of this dissertation is *to develop a modelling framework and a methodology to build holistic human-flood interaction models that provide new insights into FRM policy analysis and decision-making.* In this context, "holistic" refers to capturing both the human (i.e., communities' vulnerability and exposure including the drivers) and the physical (i.e., flood hazard) components in a coupled model using knowledge from the respective disciplines. However, it should be noted that models are abstractions of reality and could not represent all aspects of each subsystem.

To realize the aim, we formulate the following research questions:

1. *Which elements should be included to conceptualize the human-flood interaction?*
2. *How can we couple models that explicitly represent the human and the flood subsystems and the interactions between them?*
3. *How can coupled human-flood system models that incorporate institutions such as risk drivers advance FRM?*

1.6 Research approach

Based on the research gaps identified in modelling and studying human-flood interactions, in this dissertation, we investigate the merits of the CAS perspective and the integrated modelling approach to build holistic models for FRM. We first develop a modelling framework to decompose the elements that make up human-flood systems. The framework defines the coupled system as a CAS and conceptualizes the drivers of flood hazard, vulnerability and exposure as factors that shape the complex interactions between and within the component subsystems.

In the methodology that accompanies the framework, the human subsystem is modelled using the agent-based modelling approach. Consequently, the framework incorporates heterogeneous actors and their actions and interactions with the environment and flooding. It also provides the possibility to analyse the underlying institutions that govern the actions and interactions in managing flood risk. The flood subsystem is modelled using a physically-based, numerical model. The ABM is dynamically coupled to the flood model to model the interactions between the subsystems.

Then, applying coupled ABM-flood models, we investigate the effects of different institutions for FRM policy insights in two case studies. The first case study is conducted to evaluate existing and proposed FRM policies. The institutions are mainly formal ones, which are available in written documents. In this case study, we use one flood events series in all the coupled model experiments. In the second case study, the focus is on the influence of informal institutions in which individuals' behaviour to adopt measures are affected by the actions of their social network. We also test the effects of several flood events series on the flood risk mitigation behaviour of individuals.

1.7 Scope

This research is carried out within the European Commission's Seventh Framework Program Preparing for extreme and rare events in coastal regions (PEARL) project. Therefore, the research has been funded by the project. Due to project objectives and requirements, the research analyses human-flood interactions for long-term FRM measures — both public and individual measures that reduce flood hazard and communities' exposure and vulnerability. Human-flood interactions at the operational level are not addressed in this dissertation. Furthermore, the study sites selected in this research are part of the PEARL project.

1.7.1 Scientific relevance

In this dissertation, flood-related disasters are addressed as "physical disasters" instead of "natural disasters". We acknowledge that humans' role to a flood disaster is as significant as the danger from the physical event. Hence, models that help flood risk managers and other decision-makers to grasp the whole picture better and reduce flood risk need to incorporate both the human and the flood components explicitly. Despite the recent advances in socio-hydrologic modelling, such models that holistically and explicitly address the human-flood interaction for long term FRM are not available in the literature. This research fills that gap by developing a framework that defines the elements that should be considered in the coupled system and their non-linear interactions. Sivapalan (2015, p. 4800) characterizes the use of stylized socio-hydrologic models, which use differential equations to conceptualize the human-flood dynamics, as: "doing social science by natural science methods." This dissertation also fills that gap by developing a methodology to build a coupled ABM-flood model that uses domain knowledge to model human behaviour, institutions, and hydrologic and hydraulic processes that generate flood.

1.7.2 Scientific contributions

Socio-hydrology This research contributes to socio-hydrology by providing a framework to develop a model that better conceptualizes the human-flood dynamics.

The framework defines system elements that should be considered during conceptualization. It also explicitly incorporates institutions such as drivers of flood hazard and communities' vulnerability and exposure in the model conceptualization. Further, the research contributes by providing a methodology to integrate specialized modelling techniques for both the human and flood subsystems instead of "modelling social science by natural science methods."

Integration of social and hydrological sciences As it requires domain knowledge to develop coupled ABM-flood models to study human flood interactions, this research contributes towards interdisciplinary research of social and hydrological researchers. Xu *et al.* (2018) suggest that wider collaboration opportunities between the two disciplines can be achieved by introducing popular themes in the researches. In our case, bringing the agent and institution concepts into the field of hydrology/hydraulics facilitates collaboration among researchers of different background.

Flood risk management This research contributes to the FRM in multiple ways. First, it provides the holistic view of flood risk in which one could study how the social, economic, governance and hydrological makeup of an area affect the risk. The coupled model presented in this dissertation could capture all aspects of flood risk — flood hazard and communities' vulnerability and exposure. Second, the research contributes to FRM by putting emphasis on institutions and by providing policy analysis for decision makers. Coupled ABM-flood models provide a platform to test existing and proposed flood risk reduction policies. The new insights gained from simulation outputs could contribute to better FRM policy design. Finally, the coupled ABM-flood model presented in this research contributes to FRM by presenting a simulation that shows how flood risk evolves over time in response to actors' behavioural changes, measures implemented as well as environmental changes such as urbanization and climate change.

1.8 Outline

Given the research motivation, questions, objectives and scope already presented, the next six chapters are structured as follows:

Chapter 2 presents the theoretical background of the research. The main aim is framing FRM and coupledin the CAS perspective. The chapter defines and explains the main characteristics of CAS. It introduces integrated or coupled systems such as human-environment, human-water and human-flood systems. It, then, explains what agent-based and flood modelling approaches are. It also introduces institutional analysis and the meta-model we use to structure the coupled system. Finally, the chapter describes the basic steps in model integration.

Chapter 3 details the modelling framework that is developed in the research. It provides descriptions of the components of the framework and how they are related. It also presents a step-by-step method to develop a coupled ABM-flood model using the modelling framework. In this chapter, we will emphasize more on how to develop the ABM and how the coupling should be performed.

Chapter 4 and Chapter 5 present applications of the modelling framework in actual case studies to examine formal and informal institutions. In Chapter 4, the Caribbean island of Sint Maarten is selected as a case to explore the implications of mainly formal FRM institutions. Chapter 5 explores the effects of informal institutions on household adaptation measures to reduce vulnerability to flooding in the case of Wilhelmsburg, a quarter in Hamburg, Germany. Both chapters employ the framework and develop coupled ABM-flood models to simulate different FRM institutions and agent behaviours.

Chapter 6 details the insights derived from applying the CLAIM framework and the methodology such as benefits and limitations, challenges of models integration and the associated model uncertainties. It also discusses the insights gained into socio-hydrologic and FRM researches by explicitly modelling human-flood interactions using integrated models.

Finally, Chapter 7 discusses how the research outputs address the research questions. Then, it provides the personal reflections of the researcher on the modelling process and the broad experiences of the PhD journey. The chapter closes with the outlook of the research.

2

THEORETICAL BACKGROUND

2.1 INTRODUCTION

The goal of this chapter is to give definitions of key terminologies and to explain certain theories, methods and modelling approaches that will be used in this dissertation. We first discuss CAS and typical properties of such systems. Second, we explain coupled systems that are categorized as CAS. The focus will be on coupled systems designed to study human-environment interactions, human-technology interactions and human-water interactions. Third, we elaborate on two main disciplines that study human-water interactions — hydrosocial and socio-hydrology. In the section, we also emphasize on the methodologies that are implemented to study the coupled human-water systems. Finally, we focus the human-flood interactions and FRM. We will discuss the different modelling approaches implemented to study human-flood interactions, particularly flood models, ABMs, institutional analysis and integrated modelling. It should be noted that the purpose of this chapter is not to provide a systematic, detailed review of all the concepts mentioned above.

2.2 FLOOD RISK MANAGEMENT: A COMPLEX ADAPTIVE SYSTEM PERSPECTIVE

2.2.1 Complex adaptive systems

A complex system is "a system in which large networks of components with no central control and simple rules of operation give rise to complex collective behaviour that creates patterns and sophisticated information processing" (Mitchell, 2009, p. 13). The emergent behaviour of the system cannot be simply inferred from the behaviour of its components. Hence, to understand the behaviour of a complex system, one must understand not only the behaviour of its components but how those components act together to form the behaviour of the whole (Bar-Yam, 1997). Further details and illustrations on how simple programs produce complex behaviours are given by Wolfram (2002). If the system involves *adaptation* via learning or evolution (Mitchell, 2009), then it is called complex adaptive system (CAS). Adaptation is the improvement of components of a system over time in relation to the environment, which can be physical, social, technical and cultural environment (Nikolic and Kasmire, 2013).

CAS has the following common properties (Bar-Yam, 1997; Behdani, 2012; Boccara, 2004; Holland, 2014; Mitchell, 2009; Nikolic and Kasmire, 2013; Rand, 2015):

- Simple and heterogeneous components or agents that interact simultaneously — the components are considered simple relative to the whole system. The interactions occur across time, space and scale.
- Nonlinear interactions among components — there is no proportionality and no simple causality between the magnitudes of stimuli and responses, i.e., small changes in the system can have a profound effect. Thus, the whole is more than the sum of the parts.
- Information processing — perceive, communicate, process, use and produce information.
- Self-similarity or fractal-like behaviour both in structure and behaviour — as CAS is nested, higher system levels are comprised of smaller ones.
- No central control — the system organizes itself in a decentralized way.
- Emergent behaviour — the collective outcome of interactions or networks of agents which are understood on system level and not on an individual basis.
- Adaptation — the capacity to evolve based on interactions, feedback and selection pressures, and agents learn to survive or excel in their environment. Adaptation is not merely a random variation.

The main advantage of complex systems thinking is that its ability to dynamically link one part of a system (for example, a biophysical part) to another part of the same system (for example, a human part). Models which incorporate the systems thinking may consider structural change, learning and innovation and hence provide a new basis for policy exploration (Allen *et al.*, 2008). Complex systems thinking also help to fertilize cross-disciplinary integration (Bar-Yam, 1997). This integration can be achieved by developing tools for addressing the complexity of subsystem domains which can finally be adopted for more general use by recognizing their universal applicability.

Examples of CAS include ant colonies (Gordon, 2002); immune system (Ahmed and Hashish, 2006); the brain, economies, the World Wide Web (Mitchell, 2009); cities (Bettencourt, 2015); traffic, crowd movement, the spread and control of crime (Ball, 2012); ecosystems and the biosphere (Levin, 1998). Integrated or coupled systems are also categorized as CAS, and recent researches examined such systems using CAS concepts and methods. Below, we briefly describe three types of coupled CAS: coupled human and natural systems, socio-technical systems and human-water systems.

Coupled human and natural systems

Coupled human and natural systems (CHANS) are "integrated systems in which people interact with natural components" (Liu *et al.*, 2007, p. 1513). In CHANS, the human subsystem is also called social subsystem while the natural subsystem can be identified as the environment, ecology, ecosystem or landscapes[1]. Therefore, CHANS are also known as social-ecological systems (Ostrom, 2009; Schlüter *et al.*, 2012), socio-ecological systems (Filatova *et al.*, 2013), human-environmental systems (Harden, 2012) and human-landscape systems (Werner and McNamara, 2007). These systems are composed of subsystems such as resource systems, resource units, users and governance systems that interact to produce outcomes at the system level (Ostrom, 2009). As the subsystem interactions are strong, it is significant to study them as a coupled system (Werner and McNamara, 2007) through multidisciplinary efforts that address the multilevel whole system (Ostrom, 2009). Areas of focus under these systems include bio-gas infrastructures (Verhoog *et al.*, 2016), sustainable agriculture (Teschner *et al.*, 2017), land degradation (Detsis *et al.*, 2017), land use and land cover change (Drummond *et al.*, 2012), recreational fisheries (Ziegler *et al.*, 2017), coastal and marine systems (Glaser *et al.*, 2012), rangeland management (Li and Li, 2012), and wildlife conservation (Carter *et al.*, 2014).

Socio-technical systems

As CHANS are already complex, technological processes are considered as exogenous factors (Smith and Stirling, 2010). However, another class of integrated systems called socio-technical systems (STS) consider technical artefacts as an integral part of the system. STS are systems composed of two interconnected subsystems: a social system of actors and a physical system of technical artefacts (Dijkema *et al.*, 2013; Kroes *et al.*, 2006). The social system is composed of human agents and social institutions in which their interactions with the technology artefacts are embedded within complex social structures such as norms, rules, culture, organizational goals, policies and economic, legal, political and environmental elements (Ghorbani, 2013; Qureshi, 2007). Examples of STS include supply chain (Behdani, 2012), civil aviation such as aircraft maintenance (Pettersen *et al.*, 2010), wastewater treatment plant (Panebianco and Pahl-Wostl, 2006), energy systems (Bolton and Foxon, 2015; Markard *et al.*, 2016), transport system (Watson, 2012), and mobile phone production, consumption and recycling (Bollinger *et al.*, 2013).

Human-water systems

CHANS and STS are broader systems that cover wider aspects of natural resources and technical artefacts, respectively. For water managers and hydrologists, a narrower system definition that studies human-water interaction is relevant, and such system can be a coupled human-water system. As in the other coupled systems,

[1]In a coupled system, we call the components *subsystems*. The subsystems are systems by themselves but the name reflects that they are part of a bigger system. This shows that coupled systems are nested and self-similar.

the human (or social) subsystem comprises human actors and aspects such as individual and collective decision-making mechanisms and organizational structures (Blair and Buytaert, 2016). The water subsystem includes processes in the water cycle, the physical rules and water's cultural and religious significance (ibid). Studies of human-water interaction include irrigation systems (Wescoat *et al.*, 2018), water resources management (Essenfelder *et al.*, 2018), domestic water demand and use (Jepson and Brown, 2014; Koutiva and Makropoulos, 2016), FRM (Di Baldassarre *et al.*, 2013; Viero *et al.*, 2019), and water stress conditions (Kuil *et al.*, 2016).

2.2.2 Hydrosocial and socio-hydrology — sciences of human-water interaction

There are two disciplines to study the coupled human-water systems — hydrosocial and socio-hydrology[2]. Although the system descriptions in both disciplines relate humans and water components, the interactions between the two components are explained in different ways. The focuses of study and methodologies applied are different as well.

Hydrosocial

The main focus of hydrosocial is the underlying role of social power and its effect on system-level political and material inequity (Wesselink *et al.*, 2016). In this coupled system definition, water and social power are related internally, and the system analysis emphasizes on the social nature of water besides focusing on society's relationship with water (ibid).

In hydrosocial, the *hydrosocial cycle* is a fundamental concept. The hydrosocial cycle is defined as "a socio-natural process by which water and society make and remake each other over space and time" (Linton and Budds, 2014, p. 175). As such, any change in the water flow and quality through technological interventions or policy reforms (which is governed by social structure and social power) affects the social structure and power (ibid). Hence, the hydrosocial cycle internally relates entities such as social power and structures of governance, technologies, infrastructure, policies, and water itself (Linton, 2014).

Studying human-water interactions using the hydrosocial system definition is a social science perspective, mainly using human geography applications and methodologies such as historical materialist analysis (Wesselink *et al.*, 2016). For example, Akhmadiyeva and Abdullaev (2019) applied the concept of hydrosocial cycle for studying water management paradigms in the Caspian Sea region. They analysed the social (political and economic aspects), technical (technological, structural interventions) and environmental (water quality and level) dimensions of the Caspian Sea

[2]Socio-hydrology as a discipline is introduced and advocated by Sivapalan *et al.* (2012). They, in fact, call it "the new science." However, there has been quantitative human-water interaction researches that implemented modelling and simulation methods to understand underlying system behaviours and explore future trajectories before 2012.

region in different historical periods. Boelens (2014) explored interactions among water, power and cultural politics in the Peruvian Andes. The cultural and metaphysical realities of water were analysed, through local worldviews, interwoven with water flows and water control practices. The study used the metaphysical water reality construction to examine water politics and governance techniques showing water's political and social nature. Bouleau (2014) used the hydrosocial concept to examine water science and the role of scientists in defining and categorizing waterscapes in the Seine and the Rhone Rivers in France. The scientific categorization affects the political categorization of waterscapes, and the new understanding leads to institutionalizing a novel water management system. Bouleau also highlighted how waterscapes shape water science.

Socio-hydrology

On the other hand, in socio-hydrology, the main focus is studying "the co-evolution of humans and water on the landscape" (Sivapalan *et al.*, 2012, p. 1272). The notion of co-evolution underpins that the two subsystems are connected, and the actions within one evolving subsystem will have some effect on the other (Nikolic and Kasmire, 2013). The possible trajectories of the co-evolving human-water system generate a system-level emergent behaviour that gives insight to the potential future state of the system (Nikolic and Kasmire, 2013; Sivapalan *et al.*, 2012) and management strategies if needed. Hence, *interaction* is the first characteristics in socio-hydrology in which humans are an endogenous part of the water cycle, and the interaction between the two subsystems is through water consumption, pollution, policies, markets and technologies (Sivapalan *et al.*, 2012).

In addition to the direct and indirect relationships and two-way coupling between humans and water, *feedback* is crucial characteristics of socio-hydrology (Troy *et al.*, 2015a). For example, Elshafei *et al.* (2014) explained the feedback mechanisms using economic and population dynamics in relation to the use of water. Through the consumption use of available water, economic gains may increase, which further increases the population size. That, in turn, increases the water demand, and it leads to water management decisions. In response, economic gains may be limited over time and change the quantity of water. It should also be noted that exogenous and endogenous *drivers* such as people's movement (e.g., migration), market prices, climate and political situations affect the feedback (Elshafei *et al.*, 2014).

Contrary to the methodologies applied in hydrosocial studies, socio-hydrologic studies are quantitative ones that are used to test hypotheses, to model the coupled system and provide insight into the possible future trajectories of system states (Sivapalan *et al.*, 2012). Socio-hydrologic modelling of human-water systems is performed mainly from a hydrologist perspective. Sivapalan and Blöschl (2015) categorized the modelling paradigms into two types — *stylized* and *system-of-systems* models.

Stylized models are formulated mathematically using differential equations (ibid) and solved analytically or numerically. The mathematical equations are used to explicitly formalize the hypothesis about fundamental processes, the subsystems drivers, the interactions and the feedback (Di Baldassarre *et al.*, 2015). Stylized models are easy to use, transparent and able to capture the essential dynamics and emergent states of the coupled system (Di Baldassarre *et al.*, 2015; Sivapalan and Blöschl, 2015). The authors also pointed out the drawbacks of stylized models. The first one is the lumping nature of such models in which the spatial heterogeneity of human behaviour is neglected. The second is that stylized models tend to oversimplify the complexity of the coupled system. Lastly, the mathematical equations used in the models are not strongly supported by social theories.

A typical example of a stylized socio-hydrologic model can be the one presented by van Emmerik *et al.* (2014). They described the interaction and the competition for water between humans and ecosystem in the Murrumbidgee River basin, Australia using five coupled nonlinear ordinary differential equations for an irrigated area, population dynamics, hydrology, ecological/wetland water balance and environmental awareness within society. To highlight the interactions, two of the factors that govern the population dynamics are the growth or loss rate through internal relocation and the relative attractiveness level of a region. The relative attractiveness is a function of the per capita irrigation potential, and the relocation is a function of the difference in attractiveness between two regions and the environmental awareness. The irrigation dynamics is also governed by the hydrological water balance and the environmental awareness.

The system-of-systems models represent the subsystems and their components using individual models that are based on well-established methodologies from the relevant disciplines (Sivapalan and Blöschl, 2015). The authors mentioned the advantages of such models as: they are spatially explicit representing heterogeneous entities; and they represent system processes in detail. Their disadvantages are that there is an associated high cost of effort to build the models; and a realistic model parametrization is a difficult task.

An example of a system-of-systems model that couples a discrete choice model with a hydrologic model is developed by Conrad and Yates (2018) for the Okanagan Basin, Canada. The authors used a discrete choice model to estimate residents landscaping features choice based on lawn size, turfgrass variety, summer appearance and the associated water cost for the outdoor water use. They applied the hydrologic model to estimate the supply and delivery of water to residents from surface water sources. The coupled model is used to simulate lawn alternative scenarios and water users' response, and to evaluate changes in outdoor water use in five years period.

2.2.3 Human-flood systems and flood risk management

We have mentioned that human-water interaction studies cover a list of application areas such as irrigation, water resources management, domestic water demand and use, FRM, and water stress conditions. In this dissertation, we focus on FRM (the definition of FRM and other related terms is provided in Box 2.1); hence, in the coupled system perspective, we lay emphasis on the human-flood system.

Box 2.1 | Working definitions of key flood risk management terms used in this dissertation (in alphabetical order).

Flood disaster: Severe alterations in the normal functioning of a community or a society due to hazardous flood events interacting with vulnerable social conditions, leading to widespread adverse human, material, economic, or environmental effects that require an immediate emergency response to satisfy critical human needs.

Exposure: The presence of people, livelihoods, infrastructure, or economic, social, or cultural assets in places and settings that could be adversely affected by floods.

Flood: The overflowing of the normal confines of a stream or other body of water, or the accumulation of water over areas not normally submerged.

- **Coastal floods:** occur when high tides or storm surges exceed land levels or coastal defences in coastal cities or in deltas (Vojinovic and Abbott, 2012)
- **Pluvial floods:** occur when the volume of heavy rains directly falling over urban areas exceeds drainage systems capacity (Vojinovic and Abbott, 2012)

Flood hazard: The potential occurrence of a flood event or trend or impact that may cause loss of life, injury, or other health impacts, as well as damage and loss to property, infrastructure and livelihoods.

Flood risk: Probability of occurrence of hazardous flood events or trends multiplied by the impacts if these events or trends occur. Risk results from the interaction of vulnerability, exposure, and hazard.

Flood risk assessment: The qualitative and quantitative scientific estimation of flood risks.

Flood risk management: Processes for designing, implementing, and evaluating strategies, policies, and measures to improve the understanding of flood risk, foster flood risk reduction and transfer, and promote continuous improvement in flood preparedness, response, and recovery practices, with the explicit purpose of increasing human security, well-being, quality of life, and sustainable development.

Impact: Effects on natural and human systems such as lives, livelihoods, health, ecosystems, economies, societies, cultures, services, and infrastructure due to the interaction of hazardous flood events occurring within a specific period and the vulnerability of an exposed society or system.

Residual risk: The risk due to failure of technical systems, or due to a rare flood which exceeds the design flood (Plate, 2002).

Vulnerability: The propensity or predisposition to be adversely affected. Vulnerability encompasses a variety of concepts and elements, including sensitivity or susceptibility to harm and lack of capacity to cope and adapt.

Source: Unless stated, the definitions given above are based on (IPCC, 2014b). The definitions are tailored to reflect flood events only, and they are given in the context of an urban environment.

The human-flood system in which FRM is at its core is a CAS, and it satisfies CAS characteristics. Humans are heterogeneous entities that have different economic, social, psychological and political attributes. The heterogeneity could be due to an intrinsic property or something that builds up when the system evolves (Tessone, 2015). For example, in a flood-prone area, some residents may have insurance against potential flood damages on their property. In contrast, others do not have insurance because either they do not afford to pay the premium, or they think they will not be flooded.

There is also spatial heterogeneity which is characterized by topography, land use, land cover and flood extent. Humans interact with each other and their environment and with the flood subsystem. These interactions are based on institutions such as land use policies, insurance policies, emergency management guidelines and community resilience guidelines. Humans perceive, use, produce and exchange information such as flood forecast, flood maps and institutions that influence their decision.

The coupled human-flood system shows nonlinearity. Merz *et al.* (2015) pointed out that societies' response to flood scales nonlinearly to either hydrologic or economic severity of the flood. They present an example that severe flood events from the late 1980s in Germany triggered limited responses (i.e., additional flood retention basins in affected catchments), whereas a flood in 2002 led to national scale policy changes. The coupled system is self-similar or nested as well since each subsystem is a complex system made of other complex systems. For example, the human subsystem is a complex system by itself made up of complex social, economic and political systems.

System-level behaviours emerge due to the actions of heterogeneous humans and their interactions between each other, with the environment and the flood subsystem. In this study, flood risk level is the emergent behaviour in a given urban environment, and this emergent state affects individual decisions. The flood risk varies over time and space as humans learn and adapt, which can be due to feedback. Flood risk results from the interactions of flood hazard and vulnerabilities and exposures of humans and their assets. This shows that in FRM, both the flood and the human subsystem interact continuously, and there is an adaptation in response to the emergent flood risk.

Therefore, FRM can be studied using CAS models. FRM is a complex process that includes different parties and various activities that are categorized as pre-flood event prevention/mitigation and preparation, and post-flood event response and recovery (Aguirre-Ayerbe *et al.*, 2018). Pre-event activities in the preparation phase such as dissemination of flood early warning information and evacuation, and post-event activities in the response phase such as search and rescue operations happen immediately before, during or immediately after a flood event. These are operational activities and are executed for a short period. To the contrary, activities in the prevention/mitigation phases such as land use planning and construction of FRM measures, and activities in the recovery phase such as impact assessment and reconstruction takes long term planning and implementation. Our focus in this dissertation is on the long term activities.

As in the case of modelling human-water systems, socio-hydrologic modelling of human-flood systems can be implemented in stylized or system-of-systems models. Most of the system-oriented human-flood interaction studies are carried out using ABMs, and we will discuss them in Section 2.4. Examples of stylized models that conceptualize the dynamics of settled floodplains as a complex human-flood system include those discussed by Ciullo *et al.* (2017), Di Baldassarre *et al.* (2015, 2013) and Viglione *et al.* (2014). In their conceptual models, they considered hydrological, economic, political, technological and social processes co-evolve over time but can be altered by a sudden occurrence of flooding. They formalized the feedback and interactions deriving the behaviour of the system using a set of differential equations. Their conceptualization is based on societal memory or experience of prior flood events as a link between humans and flood.

Yu *et al.* (2017) also used stylized models to study human-flood interactions in the polders of coastal Bangladesh. In their conceptualization of community-managed flood protection systems, they included institutions for collective actions, in addition to societal memory, to operationalize the two-way feedback of human-flood systems. They modelled informal institutions, mainly, the norm that local people cooperate on the collective maintenance of embankments that enclose the polders because of fear of losing a good name or reputation in the community, which leads to social ostracism that outcasts defectors and refuses help in times of need.

2.3 Flood modelling

Fully understanding and managing the risks associated with flooding requires reliable modelling tools that accurately replicate flood patterns. Urban flood modelling is used to quantify the flood hazard by simulating the interactions between and within hydrological processes such as precipitation, infiltration and runoff; phenomenon such as storm surge and waves; water bodies such as rivers and seas; floodplains; and hydraulic structures such as channels, dykes and dams. It helps to establish baseline conditions regarding the flood hazard, to estimate flood depth, extent, velocity and duration, to quantify the impact on residents, properties and economy, and to explore flood reduction/mitigation measures that are suitable in the urban area of interest. It can also be used in real-time to predict a potential flood event so that through early warning, the impacts can be reduced.

Urban flood modelling is implemented using one dimensional (1D) and two dimensional (2D) hydrodynamic models. 1D models are used to simulate flows in channels and drainage pipes. The 1D shallow water flow equations are described using the set of mass and momentum conservation equations (Eq 2.1) (DHI, 2017a). Where flood flows are confined within the banks of a channel, 1D models can realistically represent the flow, and they can be used to generate results safe for decision making (Price and Vojinovic, 2008). A detailed description of 1D models, including their potential and limitations, is found in (Mark *et al.*, 2004).

$$\begin{aligned} \frac{\partial Q}{\partial x} + \frac{\partial A}{\partial t} &= q \\ \frac{\partial Q}{\partial t} + \frac{\partial}{\partial x}\left(\alpha\frac{Q^2}{A}\right) + gA\frac{\partial h}{\partial x} + \frac{gQ|Q|}{C^2AR} &= 0 \end{aligned} \tag{2.1}$$

where Q is discharge, A is flow area, q is lateral inflow, α is momentum distribution coefficient, h is water level, g is gravitational acceleration, C is Chezy resistance coefficient, R is resistance or hydraulic radius, t is time, and x is grid size.

The main limitation in 1D modelling is that the model considers only one flow direction. That means, at a given time step, the computed water level is the same along a cross-section. However, urban areas are characterized by often complex flow paths as runoff may not only be confined within a drainage system. Surface flow also occurs in urban areas and are guided by urban features such as buildings and roads layout. As a result, when large floodplains are to be included in the model, 1D schematization becomes insufficient and quite inaccurate. Hence, flow simulations over urban floodplains are better modelled using 2D hydrodynamic models.

The 2D modelling approach estimates flow depth and velocities in the x and y horizontal directions (i.e., the approach assumes vertically homogenous flow). The unsteady 2D flows are described by one mass equation and two momentum equations (in the x and y directions) as shown in Eq 2.2 (Brufau and Garcia-Navarro, 2000).

$$\begin{aligned} \frac{\partial(h)}{\partial t} + \frac{\partial(hu)}{\partial x} + \frac{\partial(hv)}{\partial y} &= 0 \\ \frac{\partial(hu)}{\partial t} + \frac{\partial(hu^2)}{\partial x} + gh\frac{\partial h}{\partial x} + \frac{\partial(huv)}{\partial y} &= gh\left(S_{0_x} - S_{f_x}\right) \\ \frac{\partial(hv)}{\partial t} + \frac{\partial(huv)}{\partial x} + gh\frac{\partial h}{\partial y} + \frac{\partial(hv^2)}{\partial y} &= gh\left(S_{0_y} - S_{f_y}\right) \end{aligned} \tag{2.2}$$

where h is the water depth, hu and hv are unit discharges, u and v are velocities, S_{0_x} and S_{0_y} are bed slopes, and S_{f_x} and S_{f_y} are friction terms along the coordinate directions.

Both 1D and 2D flows characterize flooding in urban environments. Thus, in most cases, a coupled 1D-2D modelling in which 1D drainage channel or pipe flow coupled with 2D surface flow is carried out to simulate urban flooding. There are a number of numerical modelling systems developed for commercial and research purposes. One prominent commercial hydrodynamic system is the MIKE Powered by DHI software products (https://www.mikepoweredbydhi.com/). MIKE11[3] is a 1D modelling software capable of rainfall-runoff analysis and flood routing, whereas MIKE21 is a 2D modelling software for coastal and overland flow modelling. MIKE FLOOD is a software capable of coupling MIKE11 and MIKE21.

[3]MIKE HYDRO River succeed MIKE11 starting the MIKE 2018 release.

2.4 AGENT-BASED MODELLING

The main advantage of the CAS perspective, introduced earlier, is its ability to link two different subsystems dynamically, i.e., the human subsystem and the flood subsystem, and to model their interaction. Models which incorporate the systems thinking may consider structural change, learning and innovation and hence provide a new basis for policy exploration (Allen *et al.*, 2008).

In the nested human-flood system, the human subsystem is a CAS by itself. Hence, it requires a careful selection of modelling methods to simulate heterogeneity and adaptation. For example, the classical reductionist modelling methods such as differential equations or statistical techniques such as regression and Bayesian nets have limitations in modelling CAS (Holland, 2006). These methods are characterized by restrictive or unrealistic assumptions such as linearity, homogeneity, normality, stationarity (Bankes, 2002) and tractability so that they can be solved mathematically (Gilbert and Terna, 2000; Railsback and Grimm, 2012). In addition, Holland argues that differential equations are more powerful to describe systems which can easily be approximated by linear techniques and systems that tend to reach equilibrium. It is also not easy to approximate an agent's behaviour using differential equations as the agent may have conditional actions that are governed by the rules of interaction.

Hence, methods which capture a more "realistic" view of CAS shall be used, such as *exploratory computer-based models* (Holland, 2006). These are computer programs that are used to model processes, including those with non-linear relationships (Gilbert and Terna, 2000). The emergent behaviour of the model that is described by computer programs is assessed by running the program multiple times and evaluate the effect of different input parameters. Such modelling process is called *computational modelling* or *computer simulation* (ibid).

Computer-based models provide a mental laboratory in which thought experiments can be explored to define system-level possibilities (Holland, 2006). Of these modelling techniques, ABMs[4] provide the "most natural" description and simulation of a CAS (Bonabeau, 2002), and relax the assumptions that characterize differential equations and statistical models (Bankes, 2002).

Agent-based model — Definition

An ABM is "a computational method for simulating the actions and interactions of autonomous decision-making entities in a network or system, with the aim of assessing their effects on the system as a whole" (Dawson *et al.*, 2011, p. 172). ABMs offer "a way to model social systems that are composed of agents who interact with and influence each other, learn from their experiences, and adapt their behaviours so

[4]In the following sections and chapters, we use ABM to refer to either an agent-based model or an agent-based modelling paradigm. It should also be noted that, in some literatures, agent-based models are called *agent-based simulations*, *agent-based modelling and simulations*, and *agent-based computational models*.

they are better suited to their environment" (Macal and North, 2010, p. 151). However, agent interaction is not only with each other but also with their environment (Railsback and Grimm, 2012).

ABM is used to discover the global behaviour or the emergent properties of a system based on individual agents' behaviours and interactions, providing a bottom-up modelling perspective (Nikolic and Kasmire, 2013). It is also used to study individual agents' reaction to the emergent system state (Railsback and Grimm, 2012). The emergent patterns, structures and behaviours arise through the agent interactions, but not by explicitly programmed into the models (Macal and North, 2010).

An ABM consists of three elements: a set of *agents* (an *actor* is the real "thing", and an *agent* is actor's representation in a model); set of agent relationships and methods of interaction, and agents' environment (Macal and North, 2010; Nikolic and Kasmire, 2013). An agent can be defined as "a computer system situated in some environment, and that is capable of autonomous action in this environment in order to meet its design objectives" (Jennings and Wooldridge, 1998, p. 4). The autonomy is related to agent's capability to process information and act on its own without the influence of a centralized control, make an independent decision and pursue its objective (Crooks and Heppenstall, 2012; Macal and North, 2010; Railsback and Grimm, 2012). Those authors extended some other characteristics of agents. For example, agents can be heterogeneous that they differ from each other in characteristics; agents can learn from their experience and adapt their behaviours based on the current events and, in reference to past events, to better suit to their environment; agents are social entities that interact with each other and their environment; agents have goals to achieve based on their behaviour.

Agents have *state* and *behaviour* (Jennings and Wooldridge, 1998; Nikolic and Kasmire, 2013; North and Macal, 2007). The state provides relevant information about an agent's current situation through a set of variables/attributes, and it defines what the agent is. These are information such as age, location, income and type. The state may change over time due to the agent's actions and interactions in the system dynamics. The behaviour includes the actions and interactions of the agent, and it defines what the agent does. It is influenced by agent's states, its decision making and the rules of interaction. The behaviour of agents is represented by rule-based and analytical decision-making functions (Heckbert *et al.*, 2010).

Agents may have relationship and interaction with other agents and their environment. Agents may reactively interact when they are triggered by an external stimulus, or they initiate the interaction while pursuing an objective (Crooks and Heppenstall, 2012). At the same time, agents do not interact with all agents, but they interact locally. Interactions constitute feedback between an individual agent and the external elements it interacts with (Heckbert *et al.*, 2010) that leads to change in the agent's state or behaviour by taking actions (Wilensky and Rand, 2015).

These interactions happen with respect to the rules or methods of interaction (also called *institutions* as described in Section 2.5) and agents' states and behaviours. The most common types of rules are the nested if-then-else types of decision rules (Nikolic and Kasmire, 2013). The "if" part specifies conditions, and the "then-else" part specifies the actions or decisions made by agents when the conditions are met.

The environment is the space where agents are situated in and operate (Crooks and Heppenstall, 2012; Nikolic and Kasmire, 2013). It provides the external information that an agent needs to know in addition to the structure it provides in which the agents could situate. The environment can be a continuous space, a grid cell or a social network (Crooks and Heppenstall, 2012). The environment may represent a geographical space, such as the physical features of a city using geographic information system (GIS) maps (Abdou *et al.*, 2012). In cases of such spatially explicit environment representations, agents have coordinates to show their locations, which can be static or dynamic (i.e., if agents move or not). A detailed discussion of the explicit integration and representation of space in ABMs is given in (Stanilov, 2012).

Agent-based model — Development

As in any model development, certain steps must be followed to develop an ABM. Railsback and Grimm (2012) suggest an iterative modelling cycle with six tasks: formulate the research question; assemble hypotheses for essential processes and structures that are addressed in the question; choose the model processes and structures such as scales, entities, variables and parameters, and formulate the model; implement the model by converting the verbal model descriptions to computer programs; analyse, test and revise the model; and communicate the model. Nikolic *et al.* (2013) also provide ten practical steps for developing and using an ABM, which are a more detailed version of the steps given by Railsback and Grimm.

Regarding communicating the model, Grimm *et al.* (2010, 2006) develop a standard protocol for describing ABMs. The aim of the protocol is to describe all ABMs in the same sequence so that it is easy to read and understand them. The protocol is called the *ODD protocol* based on the initials of its three blocks: Overview, Design Concepts, and Details.

- The *overview* describes the purpose of the model, its state variables and scales, conceptual description of processes and the scheduling of these processes such as how time is modelled.
- The *design concepts* provide key CAS concepts for designing, describing and understanding ABMs. The 11 design concepts are basic principles, emergence, adaptation, objectives, learning, prediction, sensing, interaction, stochasticity, collectives and observation.
- The *details* include the initialization of agent attributes and the environment, model inputs that are imposed dynamics of state variables and submodels representing the process as well as the model parametrization.

The software implementation of ABMs can be done in two ways: using all-purpose software and programming language, or using specially designed software and toolkits (Macal and North, 2010). Implementing ABMs using general programming languages such as R, Python, Java, C++ and C provides flexibility as a combination of tools and libraries can be employed. However, writing programs from scratch using these languages can be time-consuming as modellers may invest time programming non-content-specific parts such as graphical user interface, data import-export and visualization (Crooks and Castle, 2012). Other challenges are the advanced programming skills required, and the little help and support available (Nikolic *et al.*, 2013).

Therefore, modellers opt to use tools/toolkits or development environments to implement large-scale ABMs[5] (Macal and North, 2010). Toolkits are "simulation/modelling systems that provide a conceptual framework for organizing and designing agent-based models" (Crooks and Castle, 2012, p. 229). They consist of a library of pre-defined routines that modellers call to define the model. They also provide the functionality to extend their capability by integrating external libraries such as GIS libraries that provide spatial analysis and better data management.

A development/modelling environment is a "programming language or modelling suite that provides the software infrastructure for programming the agents, their states and behaviour, their interactions and the environment ... [including] support code, such as a scheduler, graph plotting, statistics collection, experiment setups, etc" (Nikolic *et al.*, 2013, p. 94). It also provides built-in functionalities to compile and execute models (Macal and North, 2010). There are also hybrid services that provide a stand-alone library and a development environment (ibid).

There are numerous ABM toolkits and development environments, and it is out of the scope of this dissertation to review all of them. A comprehensive survey can be found in (Abar *et al.*, 2017), (Kravari and Bassiliades, 2015) and (Nikolai and Madey, 2009) who review 85, 24 and 53 ABM software, respectively, using multiple evaluation criteria. Based on the benefits and limitations of ABM software given in these review papers and other literature such as (Crooks and Castle, 2012; North and Macal, 2007; Railsback *et al.*, 2006), we use the *Repast Simphony* development environment (North *et al.*, 2013) to develop ABMs presented in this dissertation[6].

Repast Simphony is a free, open source, integrated, interactive, hybrid and cross-platform Java-based ABM environment. It uses the multilingual and integrated Eclipse development environment. The most recent version is 2.7 and was released in September 2019. Repast Simphony has a high computational modelling capacity and significant model scalability level. It provides time scheduling, space management, behaviour activation, random number generation, interactive two- and three-dimensional model visualizations, GIS modelling and visualization, third-party application sets, batch runs and data collection while it is running.

It also has a high degree of support through tutorials, example models, a reference manual, frequently asked questions, Application Programming Interface (API), an active mailing list including archives and a Stack Overflow page. The main limitation

[5]In some literatures, the terminologies tools/toolkits and development environments are used ambiguously. In addition, the term "platform" is used to refer to these ABM software.

[6]Repast denotes REcursive Porous Agent Simulation Toolkit.

of the Repast Simphony environment is that it is hard to learn and requires higher development effort.

Agent-based model — Benefits and limitations

The benefits of ABM as a simulation technique include: it captures emergent phenomena that result from the interaction of agents; it provides a natural description of a system in which the model seems closer to reality; it is flexible in such a way that the modeller can tune the complexity or change levels of description and aggregation of agents (Bonabeau, 2002); it is useful to get a deeper understanding of drivers and their influence on the system characteristics and to explore various institutional arrangements and potential paths of development to assist decision and policymakers (Pyka and Grebel, 2006).

ABM also provides feedback with a visualization that allows modellers to understand and examine the system at an overall, aggregate level or an individual agent level (Wilensky and Rand, 2015). Furthermore, Axelrod (2006, p. 1568) stressed that ABM is "a wonderful way to study problems that bridge disciplinary boundaries" by addressing fundamental problems and by facilitating interdisciplinary collaboration.

The major limitation of ABMs is the difficulty of modelling human agents decision (An, 2012) due to their potentially irrational behaviour and subjective choices (Bonabeau, 2002). Modelling individual agents' behaviour and their interaction requires a description of many agent attributes and behaviour and relationships; hence, detailed data is needed to parametrize the model, and ABMs tend to have high numbers of parameters (Kelly *et al.*, 2013).

Other limitations of ABMs are the difficulty in model calibration and validation (Crooks and Heppenstall, 2012; Heckbert *et al.*, 2010), the high computational requirements associated mainly with modelling large systems (Bonabeau, 2002; Wilensky and Rand, 2015) and the low predictive power of ABMs because of their sensitivity to factors such as small variations in interaction rules (Crooks and Heppenstall, 2012).

Agent-based model Applications

The use of ABMs in FRM studies has been limited though it is gaining more attention in this decade. Researchers have been developing ABMs to investigate both operational level and strategic level flood risk reduction strategies. For example, Mustafa *et al.* (2018) used a spatial ABM and a 2D hydraulic model to investigates the impacts of spatial planning policies on future flood risk for a case study of Wallonia, Belgium. The ABM simulates urban expansion and densification in flood zones for multiple urbanization and flow discharges scenarios. Their study focused only on the elements-at-risk, especially the exposure of buildings, and do not address the vulnerability of agents.

In contrast, Sobiech (2012) developed an ABM to explore vulnerability dynamics, risk behaviour and self-protection preferences of households agents against fluvial and coastal flooding in the German North Sea Coast. Individual, relational and spatial aspects influence the agent's decisions to apply self-protection measures. Though the social vulnerability dynamics is based on empirical evidence, the model conceptualization does not explicitly capture the flood hazard and the spatial environment.

Brouwers and Boman (2010) applied an ABM to test FRM strategies in the Upper Tisza River catchment, Hungary. The agents conceptualized in the ABM are property owners, insurer and the government. Floods occur due to levee failure or seepage. The FRM strategies investigated include different government compensation rates for property owners and market-based insurance compensations in case of flood damages.

Haer *et al.* (2016) evaluated flood risk communication strategies in relations to individuals' social network and their decisions to implement measures using ABMs. Their conceptualization includes household agents, and the attitudes and decisions agents make to purchase flood insurance, use flood barriers or implement adaptation measures to reduce flood risk. The communication strategies evaluated are people-centred and a more limited top-down approach in the Rotterdam-Rijnmond region, the Netherlands.

Tonn and Guikema (2017) also used an ABM to analyse how flood protection measures, individual behaviour, and the occurrence of floods and near-miss flood events influence community flood risk. The agents in the ABM are households that also implicitly represent the community. The model conceptualization includes FRM measures such as building a dyke, elevating homes and elevating equipment; and moving out of the area based on agents risk perception and neighbours influence. The model was developed for a case study area in Fargo, North Dakota, USA.

Löwe *et al.* (2017) coupled an agent-based urban development model with a hydrodynamic flood model to assess city development, climate change impacts and flood adaptation measures. The adaptation options include a master plan controlling future urban development, reducing exposure through property buybacks, rainwater harvesting facility and increasing drainage pipe capacity. The model was tested for a pluvial flooding case in Melbourne, Australia.

Dubbelboer *et al.* (2017) developed an ABM to simulate the vulnerability of homeowners, and to facilitate an investigation of insurance mechanisms. The ABM focuses on the role of flood insurance, especially public-private partnership between the government and insurers in the UK and the re-insurance scheme Flood Re. The agents conceptualized in the ABM are homeowners, sellers and buyers, an insurer, a local government and a developer. Jenkins *et al.* (2017) utilized the ABM developed by Dubbelboer *et al.* (2017) to assess the interplay between different adaptation options; how homeowners and government could achieve risk reduction; and the role of flood insurance in the context of climate change. Both studies applied the ABM for a flood risk case of London, UK.

Liu and Lim (2018) developed an ABM to simulate a range of evacuation scenarios in flood emergencies in Brisbane, Australia. The flooding considered in the study is a fluvial one, from the Brisbane River. The agents conceptualized in the model are vehicle-based evacuees in which the evacuation is affected by departure times and communications between informed and regular evacuees.

Similarly, Dawson *et al.* (2011) estimated the likely exposure of individuals to flooding under different storm surge conditions, defence breach scenarios, flood warning times and evacuation strategies using an ABM. Their model conceptualization includes traffic simulation, risk to life in terms of exposure to certain depth and velocity of floodwater, and economic damage assessment. They modelled coastal flooding due to storm surge using a simplified raster-based inundation model.

2.5 INSTITUTIONAL ANALYSIS

As mentioned previously, human behaviour is governed by a set of rules known as institutions. Institutions are "humanly devised constraints that shape human interaction" (North, 1990, p. 3). Institutions can be expressed and modelled through institutional statements described by the ADICO grammatical syntax (Crawford and Ostrom, 1995; Ghorbani *et al.*, 2013). According to Crawford and Ostrom (1995, p. 583), "institutional statement refers to a shared linguistic constraint or opportunity that prescribes, permits, or advises actions or outcomes for actors ... Institutional statements are spoken, written, or tacitly understood in a form intelligible to actors in an empirical setting." In a way, institutions have conceptual or abstract nature, while institutional statements are linguistic statements (Basurto *et al.*, 2010).

In ADICO grammatical syntax "A" refers to attributes, "D" refers to deontic, "I" refers to aim, "C" refers to condition and "O" refers to "or else" (Crawford and Ostrom, 1995). The attribute is the actor to whom the institutional statement applies. The deontic is the modal operator which can be permitted, obliged or forbidden. The aim describes the actions or outcomes to which the institutional statement refers. It defines what action is conducted and how the action is conducted (Basurto *et al.*, 2010). The condition determines when and where the aim is permitted, obliged or forbidden. Finally, the "or else" describes the sanction for failing to comply with a rule.

If an institutional statement consists of "AIC", it is regarded as a *shared strategy*; if the statement consists of "ADIC", it is a *norm*; and if the statement contains all the five components, it is called a *rule* (Crawford and Ostrom, 1995). Scott (1995), however, classifies institutions as *regulative*, *normative* and *cognitive*. Regulative institutions are related to rule-setting, monitoring and sanctioning activities. They have legal sanctioning mechanism based on coercive power. Normative institutions are related to values and norms that define goals or objectives and the appropriate ways to pursue them. Behaviours are morally governed and are based on social obligations. Cognitive institutions are related to shared-definitions of social reality. These are culturally supported and are usually taken for granted.

Institutions can also be categorized as *formal* or *informal*. Formal institutions are "rules and procedures that are created, communicated, and enforced through channels widely accepted as official" whereas informal institutions are "socially shared rules, usually unwritten, that are created, communicated, and enforced outside of officially sanctioned channel" (Helmke and Levitsky, 2004, p. 727).

To structure and conceptualize social systems by emphasizing on institutions, and to build ABMs, the MAIA (Modelling Agent systems using Institutional Analysis) meta-model (Ghorbani *et al.*, 2013) provides a comprehensive modelling language. MAIA is a formalized representation of the Institutional Analysis and Development framework (Ostrom *et al.*, 1994), and it is the only agent-based modelling language that systematically and explicitly incorporates institutions into models. MAIA makes use of the ADICO grammar to conceptualize and model different types of institutions.

The MAIA meta-model is organized into five structures (see also Verhoog *et al.*, 2016):

1. *Social structure*: defines agents and their attributes such as properties, behaviour, the physical components they own, the information they have and their decision making criteria.

2. *Institutional structure*: defines the social context such as the role of agents and institutions that govern agents' behaviour. The ADICO syntax is used in this structure to determine the type of the institutions and agents involved.

3. *Physical structure*: defines the physical aspects of the system, such as infrastructure. It focuses on the components, compositions and connections of the physical artefacts.

4. *Operational structure*: defines the dynamics of the system. The actions executed by agents, including the conditions for the actions and the interactions, are defined in this structure.

5. *Evaluative structure*: defines the concepts that are used to validate and measure the outcomes of the system.

2.6 Integrated Modelling

Studying CAS requires understanding the social, economic, governance and physical processes, their interactions and the feedback. As a result, an *integrated assessment* of the processes in which knowledge from diverse scientific disciplines are combined, analysed, interpreted and communicated to understand better the complex phenomena is essential (Rotmans and Van Asselt, 1996). One prominent method of performing an integrated assessment for both scientific and policy analysis is by integrating expert models (ibid). Integrated environmental modelling (Laniak *et al.*, 2013) and multimodel ecologies (Bollinger *et al.*, 2015) are two examples showing the relevance and applications of model integration in complex environmental and sustainability problems.

Integrated modelling is a "method that bring[s] together diverse types information, theories and data originating from scientific areas that are different not just because they study different objects and systems, but because they are doing that in very different ways, using different languages, assumptions, scales and techniques"

(Voinov and Shugart, 2013, p. 149). In general, "the term integrated ... convey[s] a message of holistic or systems thinking ... while modeling indicates the development and/or application of computer based models" (Laniak *et al.*, 2013, p. 5). There are two ways of model integration (Voinov and Shugart, 2013).

1. *Integral models*: data from various scientific fields are collected, processed, translated into one formalism and modelled as a whole. Such models are developed commonly based on the same modelling approach.

2. *Integrated models*: already built domain models are assembled for more complex system representations. Such models are made out of two or more relatively independent components that can operate on their own and are based on different modelling approaches.

In this dissertation, we focus on the second approach as we develop integrated models to study human-flood interactions. One of the most commonly used modelling type to develop integrated models is using a *coupled component model* approach (Kelly *et al.*, 2013). These models loosely or tightly couple a process-based biophysical model (for example, a hydrodynamic flood model) with a social and economic model (for example, an ABM). The advantages of coupled component models include they explore dynamic feedback, and they may incorporate detailed representations of the studied system. The challenges of developing the models include the difficulty in conceptually and technically linking legacy models as they are developed in advance, and balancing between the complexity of component models and time and resources limitations to develop and run the models.

Although done iteratively, model integration follows five phases (Belete *et al.*, 2017):

1. Pre-integration assessment: in this phase, experts set problem statements for a study area, conceptualize the system and its components, define scenarios, define the methods of analysis, and set constraints and solution criteria.

2. Preparation of models for integration: this phase is mainly related to software engineering considerations such as selecting the programming language, model development, model modification in case of existing models, and developing wrappers for language interoperability.

3. Model orchestration: this phase is about identifying the component models that will be coupled, establishing the links between the components, defining and executing workflows, and defining the sequential or iterative exchange of data between components.

4. Data interoperability: in this phase, the main issue to address is if data exchange between component models is unambiguous, correctly mapped and translated, and the dataset is formatted in the required format.

5. Testing: this phase includes integrated model verification, model output validation and uncertainty quantification.

2.7 CONCLUSION

In this chapter, we introduced the theories, methods and modelling approaches that will be applied in the following chapters. We only focussed on the main concepts to contextualize the research in a field of study and did not systematically or critically review the literature. When discussing concepts for a specific case, we will introduce them in the respective chapters.

The research conducted and presented in this dissertation will be based on the complex adaptive system perspective. We will conceptualize the human-flood interaction and FRM as a socio-hydrologic system. Such conceptualization allows us to study the system using models by incorporating interdisciplinary knowledge from both the social sciences and hydrology/hydraulics. The framework, which we will propose in the next chapter, is mainly based on the agent, flood and institutions perspective to understand human-flood interactions. We will also present a modelling methodology in the next chapter using the integrated modelling approach. The proposed modelling paradigms are ABM and numerical flood modelling.

3

CLAIM: A COUPLED FLOOD-AGENT-INSTITUTION MODELLING FRAMEWORK[1]

3.1 INTRODUCTION

This chapter aims to propose a modelling framework that captures the main components of the coupled human-flood system and a methodology that helps to build coupled models for holistic FRM. The framework identifies relevant, common set of concepts for studying similar coupled systems uniformly. It helps to decompose and conceptualize the system, design data collection and analyse model results and general findings. A framework called Coupled fLood-Agent-Institution Modelling framework (CLAIM) integrates actors, institutions, the urban environment, hydrologic and hydrodynamic processes and external factors which affect local FRM activities. The framework defines the system as a CAS and conceptualizes the complex interaction of floods, humans and their environment as drivers of flood hazard, vulnerability and exposure.

In the methodology that accompanies the CLAIM framework, the human subsystem is modelled using ABMs. Consequently, CLAIM incorporates heterogeneous actors and their actions and interactions with the environment and flooding. It also provides the possibility to analyse the underlying institutions that govern the actions and interactions in managing flood risk by incorporating the MAIA metamodel. The flood subsystem is modelled using a physically-based, numerical model. The ABM is dynamically coupled to the flood model to understand how humans interact with the environment and to investigate the effect of different institutions and FRM policy options. We employ an integrated modelling approach to couple the two modelling approaches.

[1]This chapter is partly based on the publication: Abebe, Y.A., Ghorbani, A., Nikolic, I., Vojinovic, Z., Sanchez, A., 2019. A coupled flood-agent-institution modelling (CLAIM) framework for urban flood risk management. *Environ. Model. Softw.* 111, 483–492. DOI: https://doi.org/10.1016/j.envsoft.2018.10.015

3.2 Framework Description

CLAIM is composed of five elements: agents, institutions, urban environment, physical processes and external factors. Using CLAIM, a system can be socially and physically conceptualized and modelled as a coupled human-flood system. Such a holistic model provides the possibility to test various policy scenarios for FRM. Because of the explicit modelling and integration of such policies in the model, it is possible to explore how different scenarios affect actors and the physical environment, and vice versa. The framework also defines the system boundary and identifies the type and level of interaction within the system.

CLAIM is specifically designed for the context of urban FRM. Figure 3.1 illustrates the concepts of the framework and their relations. In the following subsections, we will describe each element by providing generic examples. But, before proceeding to describe CLAIM elements, we first provide a working definition of the human and flood subsystems in this dissertation.

- The *human subsystem* is defined as the combination of human beings (in a city scale this can also be referred to as residents), their social artefacts (i.e., social groups and ways of interactions within and between groups) and their physical artefacts such as buildings and infrastructure). It includes the social, economic, political and governance aspects of human beings.

- The *flood subsystem* is defined as the combination of hydrologic and hydrodynamic processes (e.g., precipitation, infiltration, runoff formation and routing, overland flow, storm surge and waves), technical artefacts to reduce or mitigate flood risk (e.g., drainage systems, dykes and wet/dry proofing), and the urban environment including the topography, land cover and rivers.

3.2.1 Agents

Agents represent individuals or composite actors that are a collection of actors such as an organizational entity or a household. An agent has an internal state that represents the essential variables associated with its current situation, and behaviours that relate information sensed by the agent to its decisions and actions (Macal and North, 2010). Agents' state may have intrinsic nature such as age, gender and household size. The environment may also define agents' state as agents perceive the urban environment and set their state. For example, the location and elevation of a house which can be extracted from the topographic map define the internal state of an agent. If there is a flood event, agents also perceive whether they are flooded and update their state.

The behaviour of the agent consists of its decision-making process and the action that takes place as a result. Examples of these actions include building a house, constructing FRM measures or purchasing flood insurance. Agents' behaviour can be influenced by their internal state and vice versa. For example, if there is a flood

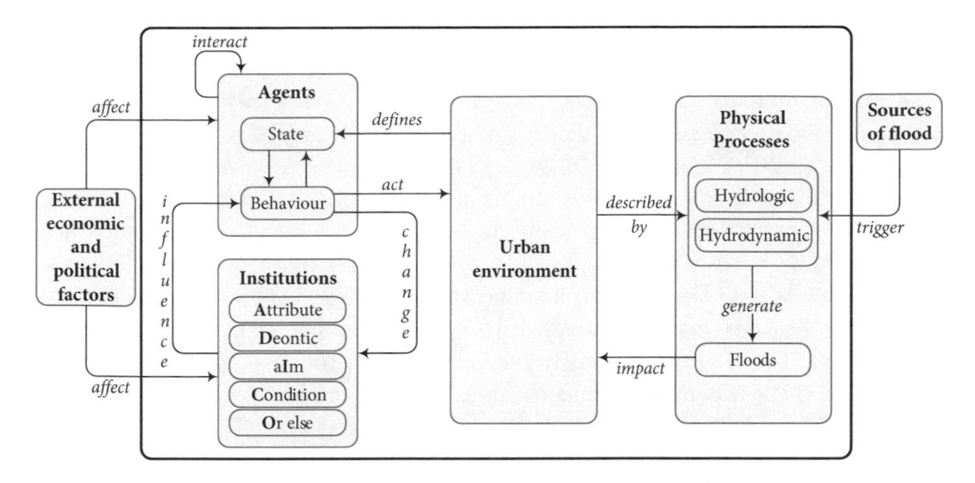

Figure 3.1 | CLAIM framework showing interactions among humans (agents and institutions), their urban environment (i.e., in the context of urban flooding), the physical processes including flood, and external factors. The drawing shows the system boundary in which elements within the outer rectangle (thick line) are related directly to local conditions and can influence each other whereas elements outside the outer rectangle affect but are not directly affected by those inside the rectangle.

event and a house is flooded (i.e., agent's state is updated), the agent may decide to protect the house by flood-proofing (i.e., the new state resulting in a change in behaviour). Alternatively, if an agent decides to build an elevated house (i.e., agent's behaviour), the house will not be flooded (i.e., agent's state remains the same) unless the flood level is higher than the floor height. As agents are social, their interactions with other agents may also change their behaviour. In the above example, an agent's decision to build an elevated house may be incentivized by an insurance firm agent through lower premiums.

3.2.2 Institutions

Humans devise institutions whose goal is to shape human behaviour. Therefore, institutions have a two-way relationship with agents in CLAIM. On the one hand, institutions may influence agents' behaviour, depending on their heterogeneity in making decisions and complying (or not) with the institutions. For example, the EU Floods Directive (European Commission, 2007) demands member states to assess the potential risk of flooding and to prepare flood hazard and risk maps. Based on these rules, member states engage in activities (i.e., influence on the government agents' behaviour) to comply with an agreed deadline.

On the other hand, agents may create, change or abolish institutions. For example, after Hurricane Sandy, the U.S. Federal Emergency Management Agency improved the map of high-risk areas for coastal flooding in New York (Dixon *et al.*, 2013). Thus, the flood insurance rate maps are also changed, which, in effect, changed the flood insurance premiums of businesses and residents.

The institutions defined here are internal (i.e., set within the system boundary) rules, norms and shared strategies that can influence agents' behaviours, and that can be changed by agents. In models, they can be defined exogenously as fixed parameters that the agents only follow or endogenously as dependent variables that are updated over time as a response to agents' behaviour. The latter may show the evolution of institutions through feedback.

In CLAIM, institutions are not directly linked with the urban environment as their impact is only through the influence they have on agents. Agents perceive and follow (or not) institutions prescribed by themselves and act on the urban environment. Conversely, agents perceive the urban environment (mainly when there is a change in the urban environment such as flooding) and may update institutions (for example, to designate floodplains as no-building zones).

3.2.3 Urban environment

Agents are situated in an environment that contains all the information external to the agent and provides space for agents' interaction (Nikolic and Kasmire, 2013). In CLAIM, agents live and build their livelihood and physical artefacts in the urban physical environment. At the same time, floods also occur in the same environment. As a result, Figure 3.1 illustrates the urban environment as a link between the human and flood subsystems. For example, if agents want to reduce flood hazard, they do not try to influence rainfall magnitude and patterns directly. They instead implement measures such as detention basins in the urban environment to retain excess rainfall.

The urban environment consists of built and natural environments. The built environment includes buildings, roads, drainage networks and flood reduction measures such as nature-based solutions, whereas the natural environment includes natural watercourses and floodplains. Changes in the urban environment are driven by the institutions and states of the agents. For example, with an increase in income level, individuals may decide to build more houses; based on a new economic policy, governments may build more roads; or to reduce recurrent riverine flooding, municipalities may invest on the construction of dykes along a river bank. As geographic information is crucial in FRM, the urban environment is a physically defined space based on GIS maps such as topography map and building and road layers. The urban environment sets the spatial boundary, and its size depends on the objective of the study.

3.2.4 Physical processes

Although the physical processes occur on the urban environment, we separate the two elements (i.e., the processes and the environment) to emphasize that our focus is only on flooding and not on other types of hazard (e.g., earthquake or landslide) that may occur in the same environment. Aspects of the urban environment that are directly linked to floods, such as drainage networks, rivers and hydraulic structures, are represented in the hydrologic and hydrodynamic processes. Depending on the magnitude of the source of flood and presence and capacity of FRM measures, flooding may occur. Flood is represented by flood maps showing its extent, depth and velocity, and the map is overlaid over the urban environment to assess the impacts on people and properties.

Agents affect the hydrologic and hydrodynamic processes through their actions on the urban environment. For example, land cover changes, such as the construction of more houses and paved parking lots, may increase the imperviousness of the surface and hence contribute to higher runoff. In contrast, implementing adaptation and mitigation measures such as green roofs, water harvesting barrels or dykes reduces runoff generation. The physical processes as well have an effect on the agents through the urban environment. Flood maps overlaid on the urban environment may define agents' internal states, for example, by changing their states from "not flooded" to "flooded".

3.2.5 External factors

There are two sets of external factors which are important in influencing the human-flood interactions: sources of flood and external economic and political factors. A flood occurs when there is a hydro-meteorological event that causes it. For example, in flash floods, the source can be intense rainfall; or in the case of coastal floods, the source can be a hurricane-induced surge. Although the hydro-meteorological events are necessary conditions for the occurrence of floods, they are classified as external factors given agents do not have the power to regulate them. Agents can only reduce the flood hazard associated with the events by implementing FRM measures (i.e., drivers of hazard).

The external economic and political factors can be institutions. Nevertheless, these factors are beyond the direct influence of the actions and interactions of agents and internal institutions in the defined urban system. Thus, in models, they can only be defined exogenously. For example, a global financial crisis may affect budgets a government agent may allocate for FRM measures. An example of external political factors can be the requirements of EU Floods Directive (European Commission, 2007) demanding member states to map and assess their flood risk.

3.3 Building models using CLAIM

As highlighted by Filatova *et al.* (2013), since ABMs primarily focus on human behaviour, integrating them with other domain modelling methods better inform policy challenges in coupled human-natural systems. Hence, to model the complex human-flood system, we use the coupled component model approach that integrates a physically-based model to model the flood subsystem and an ABM to model the human subsystem. Model integration may follow multiple phases such as pre-integration assessment, preparation of models, model orchestration, data interoperability and testing (Belete *et al.*, 2017). To build a coupled ABM-flood model, we have summarized the modelling process into four main steps:

1. Conceptualizing the system using the CLAIM framework
2. Building an ABM of the human subsystem
3. Building a flood model of the flood subsystem
4. Coupling the ABM and the flood model

Step 1 is related to the pre-integration assessment; Steps 2 and 3 are related to the preparation of component models and Step 4 incorporates orchestration and data operability.

3.3.1 Conceptualizing the system using CLAIM

Once we formulate the human-flood interaction problem that needs to be investigated, we first decompose and structure the components and processes related to the human and flood subsystems. Basically, this step is about deciding the model boundary and identifying the five components of the CLAIM framework in the coupled system. Besides guiding the collection of primary and secondary data, depending on the level of detail we want to represent in the models, this step provides the different knowledge domains or expertise required to build the agent-based and flood models.

3.3.2 Building the ABM

Considering institutions and agents' heterogeneity and adaptation, the best technique to model the human subsystem is ABM. We use the MAIA modelling language to conceptualize and structure the human subsystem and to describe it as a model formally. Agents in CLAIM, their states and behaviours, are defined in the social structure of MAIA. Agents' physical artefacts and the urban environment in CLAIM are defined in the physical structure. Institutions and the external political and economic policies in CLAIM are coded using the ADICO grammar within the institutional structure. The dynamics of the subsystem, which include agents' actions and their interactions with other agents and the environment, are defined in the operational structure.

Then, the MAIA-structured descriptions of the human subsystem are converted to pseudo-codes that can be implemented in programming languages. For the actual software implementation, the choice of modelling environments can be case-specific. One of the main criteria for choosing an ABM modelling environment is that the environment should have GIS capabilities as spatial considerations are important in CLAIM. The second criterion would be ease of use in processing results of the coupled flood model or in manipulating flood model input files.

3.3.3 Building the flood model

The flood subsystem can be modelled using a classic flood modelling technique, which couples hydrologic and hydrodynamic models. The preferred way of hydrodynamic modelling is simulating rivers and urban drainage networks (open channels or pipes) using 1D models coupled with 2D models for urban floodplains. The flood model may simulate any one or combination of fluvial, flash, pluvial, groundwater or coastal floods. Regarding flood models, any hydrodynamic flood modelling software can be used.

3.3.4 Coupling the ABM and flood model

Based on the magnitude and extent of flood hazard and their social, economic, political and governance makeup, agents may decide to implement different flood reduction and adaptation measures. To model and evaluate these measures, we couple the ABM and the flood model dynamically. In the coupled model, we consider the ABM as a "principal" model, and when we mention time steps, we are referring to the time steps of the ABM. The reason is that the ABM runs for the entire simulation period since the human subsystem dynamics take place all the time. However, since floods may not occur every time step, the flood model runs only when there is a source of flood. Moreover, for practical reasons (i.e., to be able to run the flood model automatically), the link between the ABM and flood model is embedded within the ABM. Hence, the flood model is called and executed from the ABM environment. With these considerations, we will describe how a coupled ABM-flood model can be implemented using the CLAIM framework.

As shown by the implementation flowchart in Figure 3.2, the coupled ABM-flood model method starts by initializing the agents and the urban environment. The initialization includes setting the social and physical structures of MAIA and setting up of geographic boundaries. Then, at each time step, the human subsystem dynamics, which are defined in the operational structure of MAIA, run first. These dynamics can be building houses, buying flood insurance and cleaning channels. Agents' actions and interactions that drive their exposure (e.g., building a house in higher or lower elevations, far from or close to a source of flood, and inside or outside a flood zone) and vulnerability (e.g., buying flood insurance, elevating and flood-proofing a house, or doing none of these) do not affect the flood model input files. In that case, only agents' states are updated, and the effect of the new states is evaluated later in the modelling process.

On the other hand, if their actions and interactions lead to a change in the urban environment that affects the hazard component, the hydrologic or hydrodynamic states and parameters are updated. For example, if new urban drainage networks are installed, the network file in the hydrodynamic model is updated by including the new drainage network. Since, in hydrologic models, processes are analysed per catchment, updating model parameters (e.g., imperviousness) should also be performed per catchment.

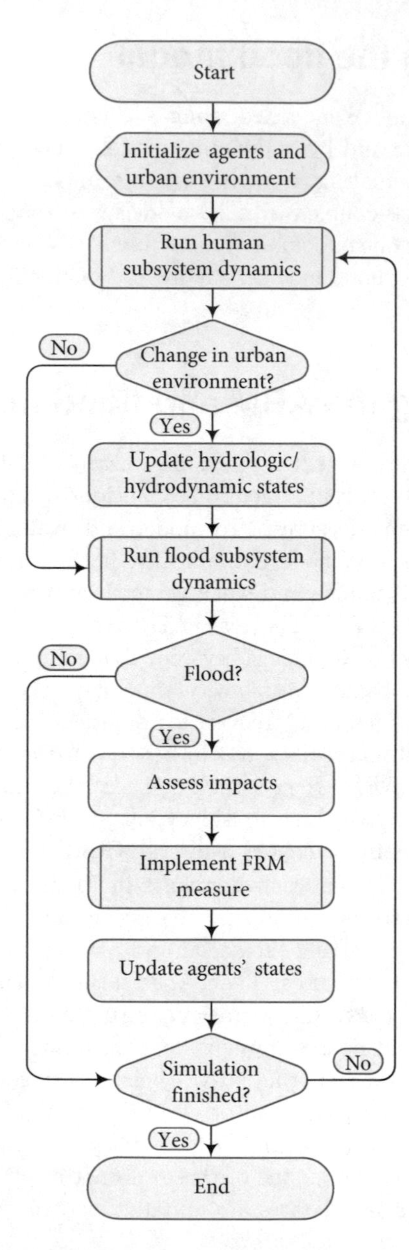

Figure 3.2 | Coupled ABM-flood model implementation flowchart for long-term FRM planning. This is a generic flowchart which should be redesigned based on case studies. The items that may differ between case studies are those mentioned in sub-process shapes (rectangles with double-struck vertical edges). These items can be expanded for specific local conditions.

Next, the flood subsystem dynamics run. As the link between the ABM and flood model is embedded within the ABM, these dynamics are also coded in the operational structure of MAIA. Such dynamics can be checking if there is a hazard triggering hydro-meteorological event, simulating the hydrologic and hydrodynamic processes, and processing model results. If there is a need to run the flood model, it must be calibrated in advance, for the first flood model run. The calibration is based on the initial urban environment setting, and after that, the flood model runs based on the continuously updated urban environment.

An essential remark here is that the implementation flowchart shown in Figure 3.2 is developed in the context of long-term FRM plans. In such a case, institutions are created/updated, or measures are implemented in a longer time scale (i.e., ABM has time steps in years) while flood events may happen for hours or days (i.e., flood models usually have time steps in seconds or minutes). Thus, we couple the two models by considering one flood event of a given duration happening within one ABM time step in which the ABM is suspended while the flood model runs. The ABM resumes once the flood model completes the run and produces the result.

Flood impacts are assessed by overlaying the flood map over the urban environment. Agents' attributes that reflect their state of exposure and vulnerability may affect the outcome of the impact assessment. For example, if the flood depth where a house is located is below the house floor level, the flood impact on that household will be zero. Since not all flood cases result in the implementation of FRM measures, the need for measures is a crucial decision-making process in FRM. Severity and frequency of flood event, the degree of impact, communities risk tolerance, availability of budget and governance are some of the factors in deciding to implement FRM measures. Moreover, where and which type of measure to implement is another dimension of the decision-making process. If a measure is implemented, the hydrologic and hydrodynamic states must again be updated.

3.4 Conclusion

In this chapter, we presented the CLAIM modelling framework, which allows for improved conceptualization and simulation of coupled human-flood systems. The human subsystem consists of heterogeneous agents and institutions which shape agents' decisions, actions and interactions, and are modelled using ABM. The flood subsystem consists of hydrologic and hydrodynamic processes which generate floods, and are modelled using numerical flood models. The dynamic link between the two subsystems happens through the urban environment.

The ABM is coupled with the flood model to study the behaviour (i.e., actions and interactions) of agents in relation to the defined institutions, and to evaluate agents' exposure and vulnerability as well as the flood hazard. The methodology presented to build a coupled model is designed considering long-term FRM plans than operational level, during-flood strategies. The output of the coupled model is a level of flood risk in terms of an assessed impact, which is used as a proxy to measure the effectiveness of the institutions in the study area.

In the following two chapters, we will apply the framework and methodology to develop coupled ABM-flood models in two case studies — the Caribbean island of Sint Maarten and Wilhelmsburg, a quarter in Hamburg, Germany. Using the models, we will investigate how formal and informal institutions affect agents' behaviour to implement flood mitigation measures, how agents influence other agents' behaviour, how floods and the environment affect agents' behaviour and vice versa. We will also test the effect of institutions on the overall flood risk of the study areas.

4

Effects of formal and informal institutions on flood risk management: The case of Sint Maarten[1]

4.1 Introduction

Hydro-meteorological and climatological disasters caused by floods, hurricanes/tropical cyclones and droughts have had damaging effects on the economies and livelihoods of populations living in small island development states (SIDS) (Mycoo and Donovan, 2017). In particular, disasters due to coastal flooding, storm surges and sea level rise pose a risk of death, injury and disruption to livelihoods (IPCC, 2014c). For example, in 2016, almost 2 million people were affected by floods in the Caribbean islands where the number of deaths reported from the floods was the second highest since 2006 (Guha-Sapir *et al.*, 2016). Housing, infrastructure and economic sectors such as tourism and agriculture also suffer from high impacts of floods.

The main reasons for the disaster impacts on small islands are that they are characterized by small land area, rapid rate of urbanization, low elevation coastal zones, concentration of human communities and infrastructure in coastal zones and high levels of informal urbanisation (IPCC, 2014b; Mycoo and Donovan, 2017; UN General Assembly, 1994). National and international initiatives were designed and implemented to reduce disaster risk recognizing the hazards, exposure and vulnerability of communities in SIDS. For example, in the Barbados Programme of Action, participating SIDS recognized the impacts of disasters and affirmed their commitment to implement national actions and policies that establish/strengthen building

[1]This chapter is based on the publication: Abebe, Y.A., Ghorbani, A., Nikolic, I., Vojinovic, Z., Sanchez, A., 2019. Flood risk management in Sint Maarten — A coupled agent-based and flood modelling method. *Journal of Environmental Management*. 248, 109317. DOI: https://doi.org/10.1016/j.jenvman.2019.109317

codes and regulatory system, promote early warning systems, establish a national disaster emergency fund, integrate disaster policies into national development plans and improve the resilience of local communities to disaster events (UN General Assembly, 1994). In addition, SIDS implement regional Comprehensive Disaster Management strategies and the Hyogo Framework for Action to address mitigation, prevention and recovery of disaster risk (DRRC, 2011). These include land use planning regulations, zoning laws, insurance funds and government contingency funds for recovery.

The challenges of implementing the policies, strategies and plans include financial constraints, lack of political commitment and lack of enforcement resulting in unregulated developments in exposed areas (DRRC, 2011). Individuals may also refuse to follow the policies as their behaviours depend on their level of risk, economic situation and awareness. This shows disaster risk reduction is the responsibility of all actors involved, from higher-level decision-makers to individuals (see also Vojinovic, 2015). Thus, in addition to quantifying the hazard, it is vital to include human behaviour and risk perception in disaster risk assessment to design relevant policies (Aerts *et al.*, 2018).

Focusing on flood disasters, we apply an ABM coupled with a numerical flood model to examine existing and proposed FRM policies in the Caribbean island of Sint Maarten. The island is selected as a case because it is frequently affected by flash, pluvial and coastal floods due to isolated rainfall events and hurricanes. Further, the Government of Sint Maarten has adopted some of the policies implemented in other SIDS, and it is planning to put in place new ones. Hence, we aim to model the FRM of Sint Maarten using coupled ABM-flood model to inform decision making and provide insights to policymakers.

Additionally, we will focus on the model evaluations, outputs analysis emphasizing on implications of institutions and agents' responses, and the resulting insights into the FRM of Sint Maarten. In the following sections, we will describe the study area, the ABM and flood models setup and input data used, model verification and validation, sensitivity and uncertainty analysis, model experimentation and results, and finally, discussions and conclusions.

4.2 Study area

The Caribbean island of Saint Martin is divided into two parts: the northern part called Saint-Martin is an overseas collectivity of France and the southern part called Sint Maarten is one of the constituent countries of the Kingdom of the Netherlands (see Figure 4.1). The study area of the present work is the island state of Sint Maarten (hereafter, the island refers to only Sint Maarten). Below, we describe the geography, climate, hydrology, as well as the organizations and institutions in the island concerning FRM.

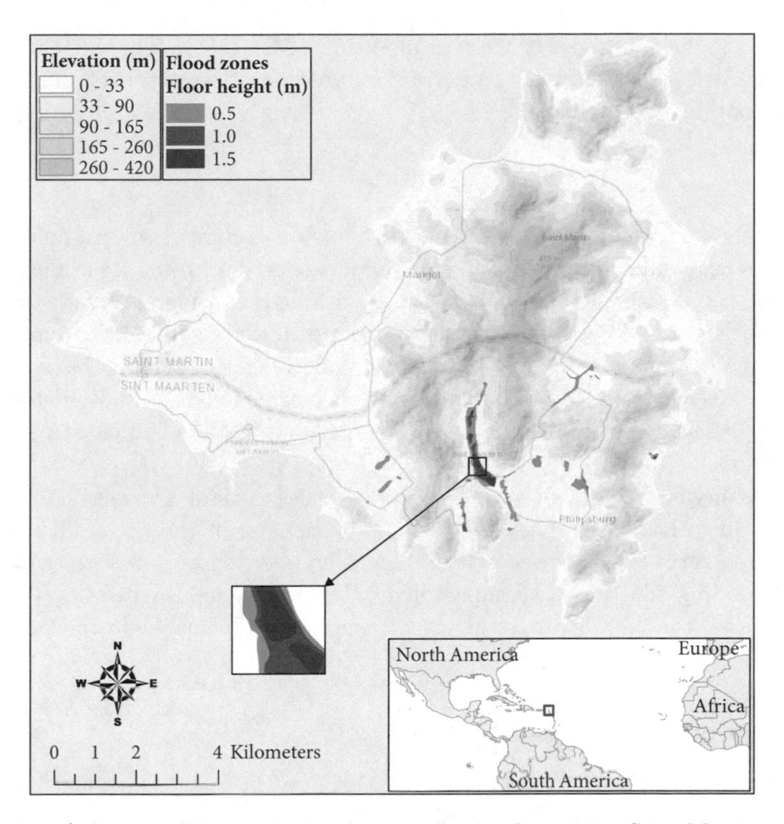

Figure 4.1 | A map of Saint Martin showing the northern part, Saint-Martin and the southern part, Sint Maarten. The map also shows the elevation ranges in the whole island. The areas in shades of red are flood zones delineated by Sint Maarten's Ministry of Public Housing, Spatial Planning, Environment and Infrastructure as part of a draft National Development Plan. New buildings constructed in the light, medium and dark red zones must have elevated floors of 0.5 m, 1.0 m and 1.5 m, respectively. (Source: the base map is an ESRI Topographic Map).

Geography and climate

The total area of Sint Maarten is approximately 34 km^2, and its total population is more than 40,000 people (STAT, 2017). The island has hilly terrains where the elevation ranges from near sea level to about 420 m above mean sea level (see Figure 4.1). The lowlands are highly urbanized with predominantly residential buildings, and businesses are situated mainly along the coast.

Sint Maarten is located within the Atlantic hurricane belt, and hence, subject to frequent hurricanes. Major hurricanes that affected the island include Hurricane Luis in 1995 (Lawrence *et al.*, 1998), Hurricanes Jose and Lenny in 1999 (Lawrence *et al.*, 2001) and, more recently, Hurricane Irma in 2017. Those hurricanes brought an enormous amount of damage to the people of Sint Maarten both economically and socially, including loss of life (see more in MDC, 2015). The damages due to hurricanes are associated with one or a combination of strong wind, storm surge, pluvial flooding and mudslides.

Hydrology

The stormwater catchments and streams in Sint Maarten have several unique characteristics that contribute to the severity of flood-related impacts. As urban environments are usually situated in low-lying areas with little consideration for stormwater drainage, they are subject to flash flooding from surrounding hills or extreme rainfall events such as thunderstorms (Vojinovic and Van Teeffelen, 2007). The stormwater channels or streams are often short, entering the ocean as low or mid-order streams. They are typically inadequate to convey runoff due to their limited capacity and obstructions.

Hurricane-induced storm surges may also cause coastal flooding. As the economy of Sint Maarten is tourism-led, many businesses in the hospitality industry are situated very close to the coastline. That increases damages and losses in case of coastal flooding. The potential impact due to hurricanes and isolated heavy rainfalls has increased considerably over the recent years with the economic and population growth on the island.

Organizations and institutions

Flood prevention, preparedness and mitigations on the island have not been sufficiently developed to cope with potential disasters. Addressing and minimizing the risk of flood-related disasters is a major challenge for the island government. For a long time, the effort of the government to manage flood risk has been concentrated on the reduction of flood hazard by canalizing and widening natural gutters and controlling stormwater levels using gates. The reasons those efforts may not always work include: the government has financial limitations to construct drainage channels in all neighbourhoods of the island; there might be a lower probability flood event beyond the channels design criteria, which can be intensified by the effects of climate change and urbanization; and gates might fail to regulate water levels during flood events.

Recently, the government acknowledges that FRM should include not only reducing the flood hazard but also reducing the exposure and vulnerability of elements-at-risk. Hence, a policy plan was drafted to improve disaster management on the island. A National Ordinance on Disaster Management is put into action to lay out the "rules and regulations" about preparation for and management of disasters, referring to immediately before and after the onset of an event.

The government is also drafting a national development plan (NDP) to manage the spatial development of Sint Maarten. This plan with zoning regulations is prepared and undertaken by the Ministry of Public Housing, Spatial Planning, Environment and Infrastructure (as commonly known as VROMI, a Dutch acronym) of the Government of Sint Maarten. The flood hazard management techniques covered in the plan are maintaining green areas, preserving and enhancing natural gutters and reserving spaces for retention ponds. The aspects specified in the plan to manage the exposure and vulnerability to flooding are the location of buildings from the sea, building codes and floor-height elevations.

4.3 Model setups

4.3.1 CLAIM decomposition

Before building the coupled model, we first decompose the coupled human-flood system of the Sint Maarten FRM into the five elements of CLAIM, i.e., agents, institutions, urban environment, physical processes and external factors. The decomposition is based on the case study description given in the previous section, the risk root cause analysis for Sint Maarten that investigates the root causes and drivers of flood risk (Fraser, 2016; Fraser *et al.*, 2016) and consultation with experts of VROMI and Disaster Management Department.

a) Agents The two main agent types considered in this case are household agents and a government agent. The household agents represent the people living in residential houses in Sint Maarten. Due to lack of data, our conceptualization does not explicitly consider businesses and public entities. However, we include the buildings these actors own by assigning them to household agents to ensure that they are considered in flood impact computation.

The government agent is a composite agent that represents the VROMI. There are three relevant departments of VROMI that the government agent represents: the Permits, the Inspection and the New Projects Departments. The first two departments are responsible for designing, regulating, inspecting and enforcing policies related to buildings, spatial planning, and development. The latter is responsible for the design and implementation of public/government buildings and drainage works. Hence, through the three departments, the government agent's actions shape the hazard and household agents' exposure and vulnerability.

b) Institutions The institutions considered are the Sint Maarten Beach policy (BP), the Sint Maarten Building and housing ordinance (BO), the Flood zoning policy (FZ) under the NDP and hazard mitigation structural measures. As beaches are an integral part of the tourism-based economy, the main objective of the BP is to protect the recreational value of the beaches on the island. The Government ensures that there is no construction of dwellings, hotels and businesses on the beach as that may restrict their normal uses. Although the policy is not formulated in relations with flood risk reduction, its implementation can have a direct effect on the exposure of household agents. Hence, it is included in the conceptualization. In the presence of natural sea sand, the policy covers up to 50 m of the strip from the coastline.

BO and FZ are drivers of the vulnerability of household agents to flood because agents are obliged to elevate the floor of new houses. The BO states that the minimum floor height of a house must be at least 0.2 m above the crown of a road whereas the FZ requires households to raise the floor of their house by 0.5 m, 1.0 m and 1.5 m as illustrated in Figure 4.1. The other difference between the two institutions is that BO is applicable to the whole island while the FZ is relevant only to delineated flood zones. It should be noted that vulnerability is a multifaceted concept (Sorg *et al.*, 2018). But, in this case, we focus only on the physical vulnerability, which is measured by the number of elevated houses.

After major flood events, the Sint Maarten Government may implement structural measures to reduce the flood hazard. The commonly implemented measures are widening channel cross-sections and constructing new ones if there is no drainage channel in the flooded area. Although it has never been implemented on the island, we also included building dykes along the coast in the conceptualization as a measure to reduce coastal flood risk.

c) Urban environment The agents mentioned above live and interact on the island. Hence, the island is part of the urban environment. Both inland and coastal floods also occur in the same environment. However, since the coastal floods are generated in a water body, we include part of the Atlantic Ocean in our conceptualization. In addition to agents, physical artefacts such as houses/buildings and drainage channels are constructed in the environment. Most houses are located in the valleys of the island though there are settlements on the hills. In some neighbourhoods where there are no drainage channels, streets drain runoff. The environment is represented by a digital terrain model as shown in Figure 4.1.

d) Physical processes The hydrologic and hydrodynamic processes included in the conceptualization are related to the inland and coastal floods. The processes include rainfall-runoff processes, 1D channel flows, 2D surface flows and hurricane-induced storm surges. Agents' dynamics such as an expansion of built-up areas and construction or widening of drainage channels on the island may affect the flood hazard by altering the imperviousness of catchments and by increasing drainage capacity, respectively.

e) External factors The sources of flooding are rainfall for inland flooding and hurricane-induced surge for coastal flooding. We do not include external political and economic factors in the conceptualization.

4.3.2 Agent-based model inputs and setup

As mentioned before, MAIA is used to conceptualize and structure the human subsystem, and provides the language to develop ABMs. The MAIA meta-model is organized into five structures: social, institutional, physical, operational and evaluative structures. Description of the first four structures of MAIA is given below.

Agents in CLAIM, their states and behaviours, are defined in the *social structure* of MAIA. The two agents defined are the household and the government agents.

- Households: Household agents make house plans, and they build houses. In the ABM environment, these agents are spatially represented by their houses. Household agents know about the three institutions, which are BP, BO and FZ, and have compliance rate attributes that reflect their behaviour to the institutions. These attributes state the level of compliance and shape agents' exposure and vulnerability. Since agents may comply with one institution but not with another, each household agent has three compliance rates corresponding to the BP, BO and FZ. These compliance rates are drawn from a uniform random distribution for every new agent at every time step.

- Government: The government agent is characterized by a level of policy enforcement. The enforcements correspond to the three institutions and are expressed using compliance threshold attributes. For BP and BO, the threshold values are set based on the percentage of houses that followed the institutions whereas for FZ, it is based on assumptions as the policy is in a draft stage. Compliance thresholds set at the beginning of a simulation are kept constant throughout that same simulation.

  In the model setup, the threshold values are expressed in percentages (or fractions) setting the rate of household agents that comply with the institution. For example, if the BP compliance threshold is 100 % (or 1), then all households need to follow the BP as agents' BP compliance rates generated from a uniform random distribution is less than or equal to 1. The government agent also constructs new drainage channels and improves existing ones to reduce flood hazard. This agent does not have a geographic representation in the ABM environment.

Agents' physical artefacts, which are the plans and the houses, and the urban environment in CLAIM are defined in the *physical structure* of MAIA.

- Plans: Before building houses, household agents develop plans to set the location, elevation and floor height of houses that will be constructed.

- Houses: The houses are also characterized by location, elevation and floor height. Houses are geographically represented by point vector data (i.e., shapefiles). Further, houses can be flooded and record their flood depth to assess the impact.

- Urban environment: The main attribute of the environment is its imperviousness, which is directly related to the number of new houses. The environment is geographically represented by a raster data of 30 m resolution.

Institutions in CLAIM are coded using the ADICO grammar within the *institutional structure* of MAIA. We code the four institutions identified during model conceptualization using ADICO as shown in Table 4.1. The BP, BO and FZ are written formal policies (although the FZ is still in draft stage) and therefore, their type is set to "rule". However, since there is no strict enforcement of the policies on the

Table 4.1 | ADICO table of institutions for the Sint Maarten FRM case.

Name	Attributes	Deontic	aIm	Conditions	Or else	Type
Beach Policy	Households	must not	build house	within 50 m of the coastline		Rule
Building Ordinance	Households	must	elevate house	regardless of location		Rule
Flood Zone Policy	Households	must	elevate house	if located in a flood zone		Rule
Flood hazard reduction	Government		implements flood hazard reduction measure	e.g., if number of flooded houses > a threshold		Shared strategy

island, there are no proper sanctions for violating these rules. Hence, the "or else" component of the ADICO is left blank.

The flood hazard reduction, on the other hand, is of a type "shared strategy" that is implemented by the government agent to reduce flood risk. As there is a budget constraint to implement flood reduction measures in every flooded neighbourhood in Sint Maarten, the government gives priorities based on the number of flood houses in a hydrologic catchment (i.e., based on flood model outputs).

The dynamics of the human subsystem, which include agents' actions and their interactions with other agents and the environment, are defined in the *operational structure* of MAIA. Figure 4.2(a) shows all the actions and interactions conceptualized in the coupled model. In the Sint Maarten ABM, we define two agents' dynamics: urban building development and FRM (flood hazard reduction).

- Urban building development: In our conceptualization, a new house is built (or planned) when there are new household agents as all households are represented spatially by a house they live in. We simplified the housing expansion mechanism in which the number and locations of new houses are based on the building permits issued by VROMI and on the NDP land use map. That is, new agents choose from a predefined set of potential future house locations randomly.

  Every time step, household agents make house plans by deciding where to build new houses and if they elevate the floor height of the houses. For example, if an agent develops a plan to build a new house, the first institution the agent checks is the BP (see Figure 4.2(b)). If the planned house's location is within 50 m from the coastline and the agent's BP compliance rate is less than the BP compliance threshold, the agent complies with BP. In such a case, the plan will be cancelled, and the planned house will not be built. But, if the agent's BP compliance rate is greater than the threshold, the agent will build the house within 50 m of the coastline.

If the plan is not cancelled, the next institution the agent checks is the FZ (see Figure 4.2(c)). If the planned house is located in the flood zones and the agent's FZ compliance rate is less than the FZ compliance threshold, the agent complies with the FZ and the house plan will be improved to change the floor elevation to the height stated in the policy. In that case, since the minimum floor elevation in FZ (i.e., 0.5 m) is higher than the floor elevation stated in BO (i.e., 0.2 m), there is no need to check the compliance to the BO. But, if the agent does not comply with the FZ, the agent will check if it complies with the BO (see Figure 4.2(d)).

Similarly, if the agent's BO compliance rate is less than the BO compliance threshold, the agent complies with the ordinance and the house plan will be improved to change the floor elevation to 0.2 m. If the BO compliance rate is greater than the threshold, the house will not be elevated. The newly built house will have the same location, elevation and floor height as the plan.

- FRM (flood hazard reduction): in Sint Maarten, most flood hazard reduction measures are implemented in a reactive, ad hoc manner. There is no systematic way of prioritizing neighbourhoods that are frequently flooded. When the budget for the construction of measures comes from the government, neighbourhoods may be selected based on their political alliance (for example, campaign promises during elections). In case budget comes from donor funds, priority may be given to economically poor areas (for example, to improve sanitation and drainage). As a result, the dynamics only run if there is a flood event.

  In the model, the government agent selects a maximum of one catchment at a given time step where a measure is implemented based on certain conditions (Figure 4.2(e)). The first set of conditions checked are: (i) if there is a rainfall event with a recurrence interval of 50-year or above as these magnitudes of rainfall causes major flood, or (ii) if a previous measure was implemented at least three years before the "current" time step, assuming it takes an average of three years to implement a measure and all relevant budget is directed to that measure. When those criteria are met, the next set of conditions are if the number of flooded houses in a catchment is greater than a threshold, and if that number is the highest.

After structuring the system using the MAIA meta-model, the descriptions and flowcharts are converted to pseudo-codes that can be coded in object-oriented programming languages. The ABM is implemented using the Java-based Repast Simphony modelling environment (North *et al.*, 2013). The environment is selected as it provides capabilities for spatial data analysis, and the Java programming language it uses provides ease of integration of the ABM and flood model (for example, in terms of input-output data processing). The full ABM software, together with the ODD protocol (Grimm *et al.*, 2010) can be accessed at https://github.com/yaredo77/Coupled_ABM-Flood_Model. Model assumptions that have been made during the conceptualization are listed in Appendix A.

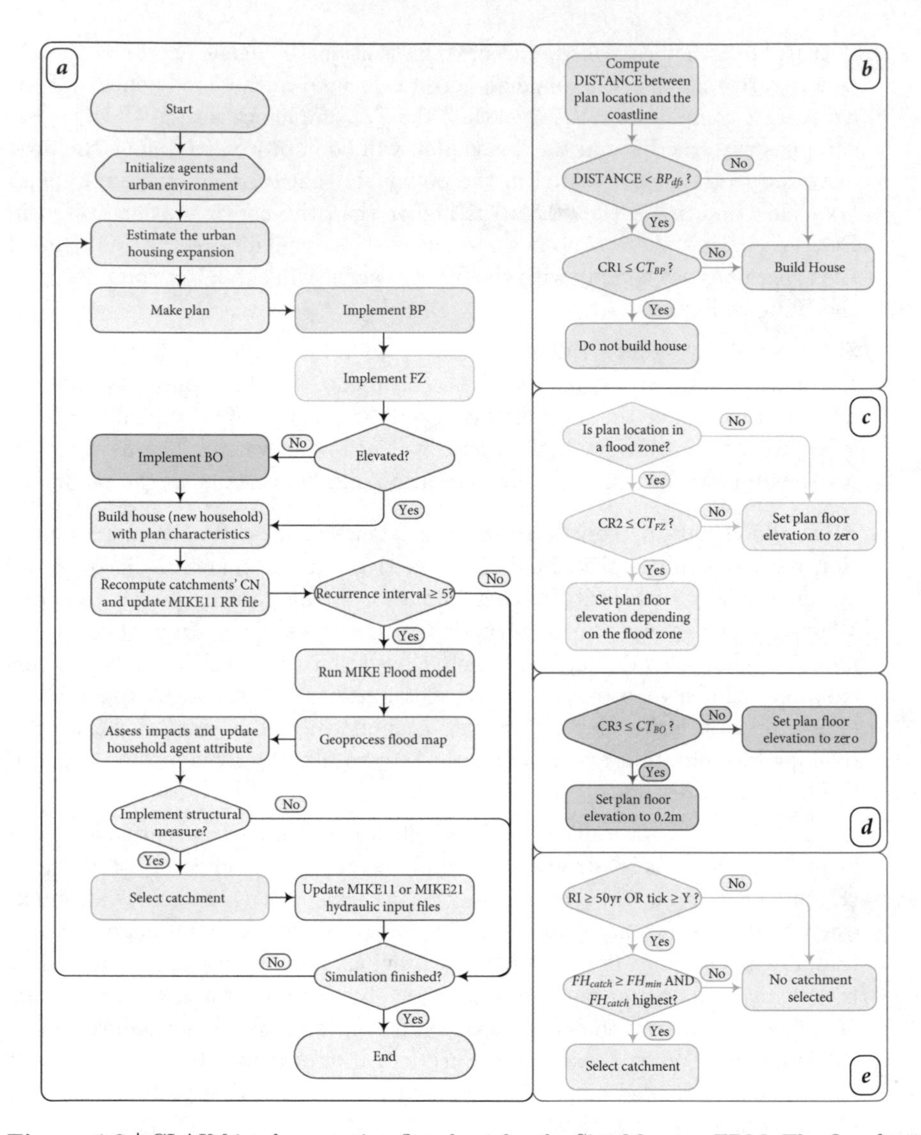

Figure 4.2 | CLAIM implementation flowchart for the Sint Maarten FRM. The flowchart also shows the coupling process (i.e., ABM-flood model coupling) as the coupling is done from the ABM modelling environment. (a) shows the general flow chart while (b), (c) and (d) show how the BP, FZ and BO policies are implemented, respectively. (e) shows the criteria to select catchments where structural measures are implemented. In the figure, CN is curve number; RR is rainfall-runoff; CR is compliance rate, RI is recurrence interval, tick is the ABM time step, Y is the years between the implementation of consecutive measures, BP_dfs is the BP distance from the sea, CT_{BP}, CT_{FZ}, and CT_{BO} are the compliance thresholds for BP, FZ and BO, respectively, and FH_{catch} and FH_{min} are the catchment and minimum (threshold) number of flooded houses, respectively.

4.3.3 Flood model inputs and setup

In the flood model, we consider both pluvial and coastal floods. The inflow for the pluvial flood simulations comes from design rainfall events of 5-year, 10-year, 20-year, 50-year and 100-year recurrence intervals. The maximum intensities of these rainfalls are 52 mm/h, 62 mm/h, 76 mm/h, 90 mm/h and 100 mm/h, respectively. The island is divided into sub-catchments, and the rainfall-runoff process of each sub-catchment is analysed with the unit hydrograph method (DHI, 2017a).

In this method, the excess rainfall is calculated by the runoff curve number (CN) method. The factors that determine the CN values of a catchment include the soil type, land cover, treatment, hydrologic condition, size of impervious areas and the antecedent moisture condition at the start of the storm. In this study, the CN values are updated depending only on the increase in impervious surfaces due to the urban developments defined in the operational structure of MAIA.

The inflow for the coastal flood simulations comes from open boundaries in which a boundary condition of 0.5 m water level is used. The value is derived from a hurricane storm surge simulation and it is the same in all flood simulations. In contrast, the inflow for the pluvial flood comes from a 1D runoff routing. The model bathymetry used in the 2D, for the pluvial and coastal simulations, has a spatial resolution of 30 m (shown in Figure 4.3).

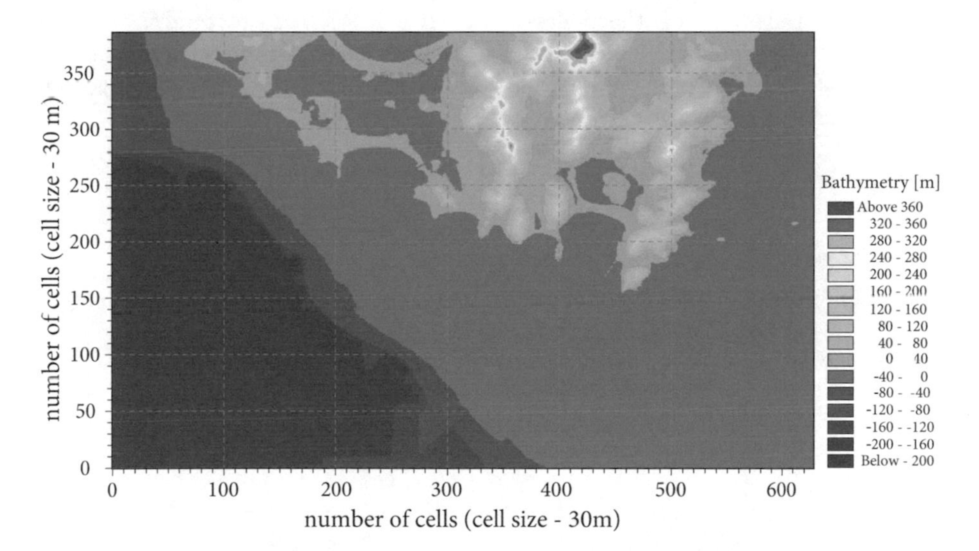

Figure 4.3 | Bathymetry used in MIKE21 coastal and pluvial flooding simulations (based on Vojinovic *et al.*, 2013). The model domain is 18.8 km by 11.6 km with a grid resolution of 30 m. This is also the same urban environment used in the ABM.

The flood subsystem is modelled using the MIKE FLOOD hydrodynamic modelling package, which couples MIKE11 and MIKE21 (DHI, 2017b). MIKE11 is used to model the rainfall-runoff processes, and 1D flows in the drainage channels while

MIKE21 is used to model the 2D coastal and pluvial flooding in the urban floodplains. The output of the MIKE FLOOD model is a map showing the flood extent and depth.

4.3.4 Coupled model inputs and setup

The ABM and flood model are coupled within the Repast Simphony ABM environment so that we use one programming language, and it is suitable to manage the input-output data of the two models. Hence, we conceptualize the coupling within the operational structure of MAIA. Figure 4.2(a) shows the coupling process. The computation time step of the ABM is one year as it takes years to build houses and flood hazard reduction measures. As the urban development agent dynamics happen at every time step, the ABM runs during the whole simulation period. However, since flooding does not occur every year, the flood model does not run every time step. The coupled model computation time step is the same as the ABM time step. When there is a rainfall event in a given time step, the flood model runs on a different timescale.

In all the simulation runs, we use the synthetic design rainfall event series shown in Figure 4.4. The input parameters and variables used in the coupled model and the ABM together with their default values are presented in Table 4.2. In the table, the input parameters and the policy-related fixed variables are *control variables* whereas the other policy related scenarios that are used to set up experiments are *independent variables* (see Lorscheid *et al.*, 2012, p. 29-30 for definitions of dependent, independent and control variables).

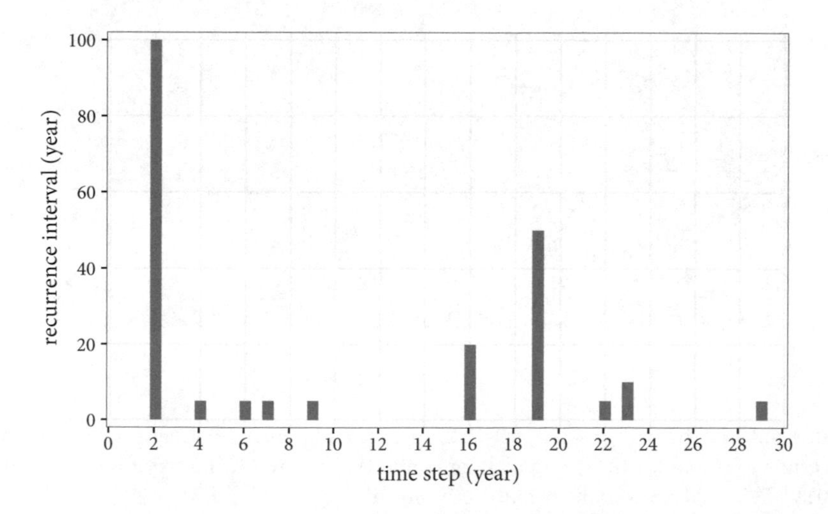

Figure 4.4 | Input design rainfall events series. It shows discrete recurrence intervals in years assuming that there is a maximum of one major flood event per time step. The coupled model runs for 30 years of simulation period in which the flood model runs ten times.

Table 4.2 | Base values of input parameters and variables used in the coupled model. The fixed policy-related variables are well-defined values used in all simulations, whereas the rest of policy-related variables are used for scenario experimentation purpose.

	Input parameter or variable	**Symbol**	**Base values**	**Remark**
Input parameters	Initial number of households/houses	HH_{ini}	12000	Based on a buildings shapefile (ca. 2010) obtained from VROMI
	Minimum number of flooded houses in a catchment that triggers structural measure	FH_{min}	20	Authors estimation[a]
	Number of years between the implementation of consecutive structural measure	Y	3	Authors estimation[a] This is only when a flood event is caused by a rainfall of recurrence interval less than 50-year
	Increase in CN of catchments per house	CN_h	0.1	Authors estimation based on 0.02ha average lot size per house
	Initial number of houses with elevated floor (BO)	$HElev_{ini}$	$0.8HH_{ini}$	Authors estimation[a]
Policy-related variables (fixed)	Floor height elevation (BO)	FH_{BO}	0.2 m	Building and Housing Ordinance, February 2013
	Floor height elevation (FZ)	FH_{FZ}	0.5 m, 1.0 m and 1.5 m (depending on the zone)	Draft Sint Maarten National Development Plan, 2014
Policy-related variables (to set up experiments)	BP compliance threshold	CT_{BP}	75 %	Authors estimation[a]
	FZ compliance threshold	CT_{FZ}	75 %	Authors estimation[a]
	BO compliance threshold	CT_{BO}	80 %	Authors estimation[a]
	Structural measures	SM	No	Authors estimation[a]
	Beach policy distance from the sea	BP_{dfs}	50 m	Sint Maarten Beach Policy, August 1994

[a] These estimations are based on discussions with Sint Maarten Disaster Management and VROMI experts.

4.3.5 Model evaluation

Model verification and validation

The flood model is developed using the commercial off-the-shelf software MIKE FLOOD, MIKE21 and MIKE11. Hence, we rather focus on verifying the ABM and the coupled model computer programs we developed. To verify the ABM, we use the *evaluative structure* of MAIA, which indicates the relationship between expected outcomes and agent actions. We record, debug and assess selected evaluation variables and check whether their values match agents' actions. For example, there is a direct relationship between the number of elevated houses and the compliance thresholds of FZ and BO. If CT_{FZ} and CT_{BO} increase, we expect to record a higher number of elevated houses. However, there is no relationship between the number of elevated houses and the BP. As another example, the number of flooded houses is directly related to the implementation of all the institutions — BP, BO, FZ and structural measures.

In the case of the coupled model verification, we monitor whether catchments CN values are updated properly reflecting the urban development in the catchments. We also monitor if the right flood map is used to compute the number of flooded houses. If there is no structural measure implementation, we expect a higher number of flooded houses when the rainfall recurrence interval at a given time step is higher. In addition, if a structural measure is implemented in a catchment, we expect to record a lower number of flooded houses in that catchment, not in any other catchment.

The flood model is validated using a historic flood event in Sint Maarten. The hydrodynamic model results, flood depth and extent, were validated against eyewitness accounts. However, it should be noted that in this study, we use design rainfall event series rather than historical flood events. Due to the lack of empirical data regarding the flood and human dynamics at the same time, validating the ABM and coupled model is a challenge.

As a result, we opted to validate the models using domain experts/problem owners from the Sint Maarten Disaster Management and the VROMI. We consulted with these experts throughout the model development process to validate the conceptualization, the modeller's estimation of input data and the model outputs. For example, earlier versions of the coupled model resulted in an overestimated number of flooded houses. The result is improved to a "reasonable" value after the experts suggested to adjust the default initial conditions of model inputs (for example, the initial number of elevated houses) and policy compliance thresholds. Given the aim of developing the coupled model is to provide insights into the long-term FRM dynamics of Sint Maarten, we do not strive to reproduce an empirically observed behaviour and system states.

Model uncertainty and sensitivity

Models are abstractions of the reality, make use of assumptions, have parameters and have initial and boundary conditions that cannot be measured/known of full certainty. It is important to perform uncertainty analysis (UA) and sensitivity analysis (SA) to understand better and communicate the outputs that inform policy decision making. In this case, we mainly focus on quantifying the uncertainty and sensitivity of the ABM and the coupling process. Although computationally intensive, recent ABM studies analyse uncertainty and sensitivity using Monte Carlo simulations that are based on samples of the full range of model input factors (Fonoberova *et al.*, 2013; Ligmann-Zielinska *et al.*, 2014). Those studies employ the global SA approach based on variance decomposition (Saltelli *et al.*, 2008).

In another study, ten Broeke *et al.* (2016) conclude that global SA does not adequately address issues such as nonlinear interactions and feedback, and emergent properties in ABMs. They recommend using one-factor-at-a-time (OFAT) SA, by varying one parameter at a time while keeping all other parameters fixed, to address the issues mentioned above. Then, they recommend performing global SA methods to address interaction effects of parameters.

Considering the computational cost of performing both OFAT and global SA, in this study, we only perform the OFAT SA analysis. To further reduce the computational cost associated with this analysis, we first performed an initial UA of the coupled ABM-flood model output with respect to the uncertainty of the 2D flood model computational grid. High-resolution topography data may provide a better representation of urban features in urban flood modelling. However, using high resolution computational grids in 2D flood modelling is computationally demanding. This implies that performing SA and UA for the coupled ABM-flood model that uses high-resolution topography data significantly increases the computational cost.

The initial UA evaluates the effect of 10 m, 30 m and 60 m computational 2D grids on the total number of flooded houses. The simulations are carried out using the default input parameters and variables set in Table 4.2. Each simulation is replicated 20 times, considering the stochasticity of the ABM that is caused by the randomization of household agents' compliance rates. All simulations in this study are performed using the SURFsara high performance computing cloud facility (https://userinfo.surfsara.nl/systems/hpc-cloud) with Windows 64x operating system and 9 CPUs.

To perform the SA, not all the model input parameters and variables are selected. The fixed policy-related variables (see Table 4.2), FH_{BO} and FH_{FZ}, are formally defined values recorded in ordinance/policy documents. The other policy-related variables are the independent variables used to set up experiments that test the effect of agents' behaviours on institutions and how that affect the overall flood risk in Sint Maarten. Hence, all the policy-related variables are set to their default values in the SA. The input factors selected for the SA are the five control variables listed in Table 4.3.

We perform the OFAT SA for the input factors specified in Table 4.3. In each simulation, we run the model for the extreme values of the input factor range and four equidistant points in between; hence, six runs per factor. We run 20 replications per factor setting to show the uncertainty of the coupled model output. The

Table 4.3 | Selected input factors for the sensitivity analysis and their uniform distribution bounds.

Simulation	Input factor	Distribution	Range
1	HH_{ini}	Uniform	[10000, 12500]
2	$HElev_{ini}$	Uniform	[0.5, 1]
3	CN_h	Uniform	[0, 0.5]
4	FH_{min}	Uniform	[10, 50]
5	Y	Uniform	[1, 6]

first three simulations are executed in a case where no hazard reduction structural measure is implemented whereas for Simulations 4 and 5 (in Table 4.3), measures are implemented.

4.3.6 Experimental setup

To assess the effect of institutions on the hazard, vulnerability and exposure, we run simulations by varying the values of the policy-related variables (see Table 4.2). The BP-related variables, CT_{BP} and BP_{dfs}, affects the exposure, whereas the FZ- and BO-related variables, CT_{FZ} and CT_{BO}, influences the vulnerability of agents. Whether there is a structural measure or not, i.e., the value of SM, affects the level of hazard.

As shown in Table 4.4, the compliance threshold values for the FZ and BP ranges between 0 and 1 to test the extreme conditions of no compliance/no enforcement and total compliance/full enforcement, respectively. In the case of the BO compliance threshold, the lower value is set to 0.5 because many houses in Sint Maarten are already elevated.

We also tested the effect of the BP buffer zone that prohibits the construction of buildings. In addition to the default value of 50 m, we test no-building zone of 0 m, and 100 m from the coastline. Finally, the scenarios for the implementation of

Table 4.4 | Policy-related variables and their value range used in the experimental setup.

Scenario variable	Value range	Step
CT_{BP}	[0, 1]	0.25
CT_{FZ}	[0, 1]	0.25
CT_{BO}	[0.5, 1]	0.25
SM	No or Yes	—
BP_{dfs}	[0, 100] in m	50 m

structural measures are based on the Boolean values of Yes and No. For the other input factors, we used their default values, as stated in Table 4.2. The rainfall event series used for all the scenarios is the one shown in Figure 4.4.

4.4 Results

4.4.1 Uncertainty and sensitivity analysis

The approximate computation time a single simulation of coupled ABM-flood model takes to run using 10 m, 30 m and 60 m 2D grid sizes is 120 h, 6 h and 1.5 h, respectively. A single simulation has 30 time steps in which the 2D flood model runs in 10 of the time steps as shown in Figure 4.4. Although using the 60 m 2D grid reduces the computational time, Figure 4.5 shows that the total number of flooded houses are lower compared to the results when using the 10 m and 30 m 2D grids, especially during rainfall events with higher recurrence intervals. This can be due to the shallower flood depths associated with the low-resolution grid (see Vojinovic *et al.*, 2011), and the floor elevations as a result of complying with BO and FZ are greater than the flood depth.

On the other hand, in most cases, the differences in the total number of flooded houses when using the 10 m and 30 m 2D grids are within the simulation output distributions as illustrated by the boxplot in Figure 4.5. However, running simulations using the 10 m grid requires 20 times the computational time required to run sim-

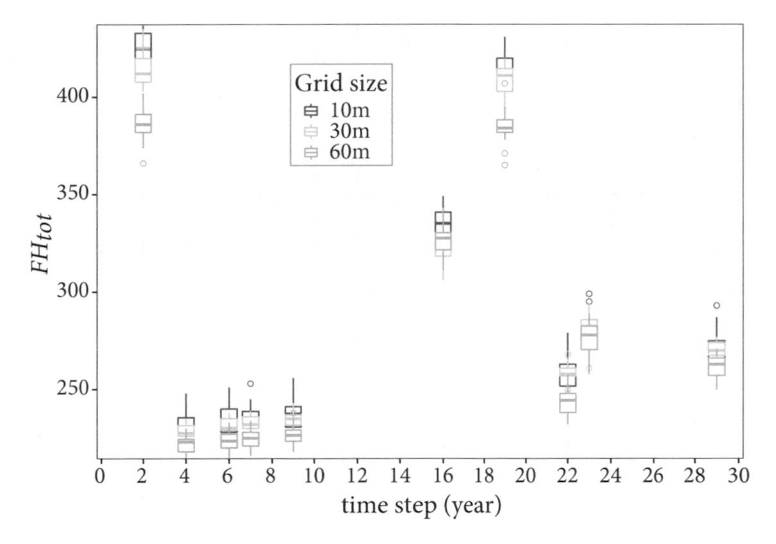

Figure 4.5 | Coupled ABM-flood model simulation outputs — total number of flooded houses (FH_{tot}) — when using 10 m, 30 m and 60 m computational grids in the 2D flood model. Each boxplot corresponds to 20 replicated simulations. The distributions are the result of the stochasticity of the ABM as heterogeneous agent behaviours are generated randomly.

ulations using the 30 m grid. Hence, in the subsequent uncertainty and sensitivity analysis and the scenario experiments, we use the 30 m 2D computational grid.

The SA results in Figure 4.6 show that all factors but $HElev_{ini}$ have a direct relationship with the number of flooded houses. Increasing the initial number of household agents increases the number of exposed and vulnerable houses, which in turn increases the number of flooded houses. Higher CN_h intensifies the flood hazard while higher FH_{min} and Y reduce the chance of structural measures implementation, increasing the flood impact. Increasing $HElev_{ini}$, in contrast, reduces the vulnerability of household agents, resulting in lower flood impact.

Based on Figure 4.6, the important factors are HH_{ini}, $HElev_{ini}$ and CN_h. The first two factors show a uniform relationship between the range of the factors and the range of the median number of flooded houses. Hence, these factors exhibit a

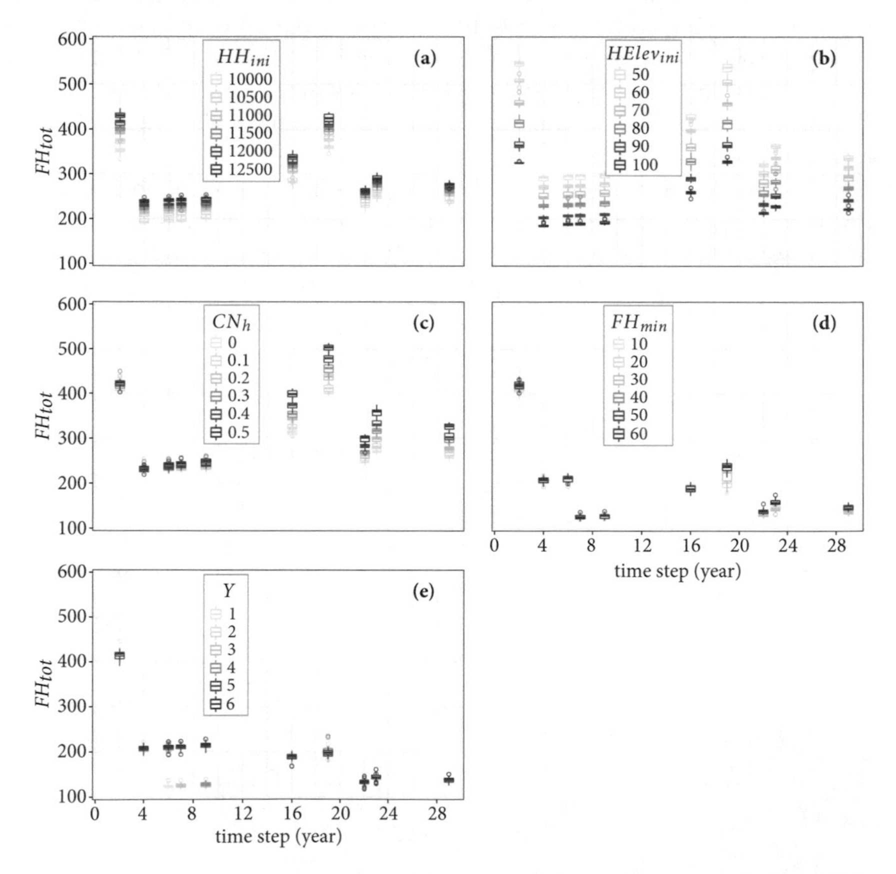

Figure 4.6 | OFAT sensitivity result for (a) initial number of households (HH_{ini}), (b) initial number of houses with elevated floor ($HElev_{ini}$), (c) increase in CN of catchments per house (CN_h), (d) minimum number of flooded houses in a catchment that triggers structural measure (FH_{min}) and (e) number of years between consecutive structural measures (Y). FH_{tot} is the total number of flooded houses.

linear relationship with the total number of flooded houses. But, the change in the value of CN_h is more important towards the end of the simulation time when more houses are built. Of the three factors, $HElev_{ini}$ is the most crucial factor.

FH_{min} and Y have a marginal effect on the total number of flooded houses. The latter has more effect in the first half of the simulation time, but its effect diminishes in the last half. After *time step* $= 9$, the next flood happens seven years later. Therefore, there is an implementation of a measure as the maximum Y in the SA is six. The other reason is that structural measure implementation is not only dependent on Y (see the illustration in Figure 4.2(e)). As there is a 50-year event at *time step* $= 19$, a measure is implemented irrespective of the value of Y.

In summary, the UA shows the underlying uncertainty embedded in the bathymetry input data. The analysis justifies why a 30 m grid bathymetry is used in the flood model. The SA analysis highlights that the total number of flooded houses is sensitive to HH_{ini}, $HElev_{ini}$ and CN_h. Model result interpretations, discussions and conclusions presented in the following sections are subject to the uncertainty and sensitivity of input factors discussed above.

4.4.2 Experimentation results

The effect of the Beach Policy on the exposure of houses

As the BP prohibits the construction of buildings within a certain distance from the coastline, it directly affects the exposure component of the flood risk. That means, if households do not follow the BP or if there is no strict enforcement of the policy, more buildings will be constructed on coasts exposed to potential coastal flooding. Figure 4.7(a) shows the worst case scenario when BP_{dfs} is zero, which effectively means there is no policy. In that case, there is no violation of or no need of enforcing a policy. Hence, despite the value of the CT_{BP}, the cumulative number of households (HH_{cum}) that do not follow the BP is also zero.

Figure 4.7(c) and (e) show HH_{cum} that do not follow the BP when BP_{dfs} is 50 m and 100 m, respectively. In these cases, since there is a policy, there can be violations based on the value of the CT_{BP}. The figures show that HH_{cum} that do not follow the BP decreases when the CT_{BP} increases. However, the number of exposed houses shows major reduction between the CT_{BP} values of 0.5 and 1 than between 0 and 0.5. For example, HH_{cum} that do not follow the BP reduces by about 36 % when the compliance threshold increases from 0.5 to 0.75 compared to only about 5 % reduction when the compliance threshold increases from 0 to 0.25. This shows that starting from zero, it seems little effort of complying or enforcement does not payoff, but through time and with more effort of complying or enforcement, the payoff increases as more households follow the BP.

Regarding widening the no-building zone, increasing the BP_{dfs} value from 0 m to 100 m results in an increase in the number of potentially exposed people. This is because more households can be affected by widening the no-building zone. For example, the maximum number of affected households increases from about 120 to 350 units with an increase in BP_{dfs} from 50 m to 100 m. That means, with more compliance or enforcement (i.e., higher CT_{BP} values), the number of exposed households will show a major reduction when the no-building zone widens.

The effect of the BP, in terms of increasing the values of CT_{BP} and BP_{dfs}, on the total number of flooded houses (FH_{tot}) is very little. Figure 4.7(b), (d) and (f) show that when the compliance threshold increases, there is a marginal reduction in the median FH_{tot}, especially towards the end of the simulation period. That is more visible when the no-building zone widens. For example, at time *time step* = 29, the increase in CT_{BP} from 0 to 1 has almost no contribution in reducing the total number of flooded houses when BP_{dfs} is zero while it contributed about 10 % reduction when BP_{dfs} is 100 m. The reason is that the BP affects a small group of agents along the coast and some part of the Sint Maarten coast is a cliff that is higher than the storm surge level simulated in the flood model.

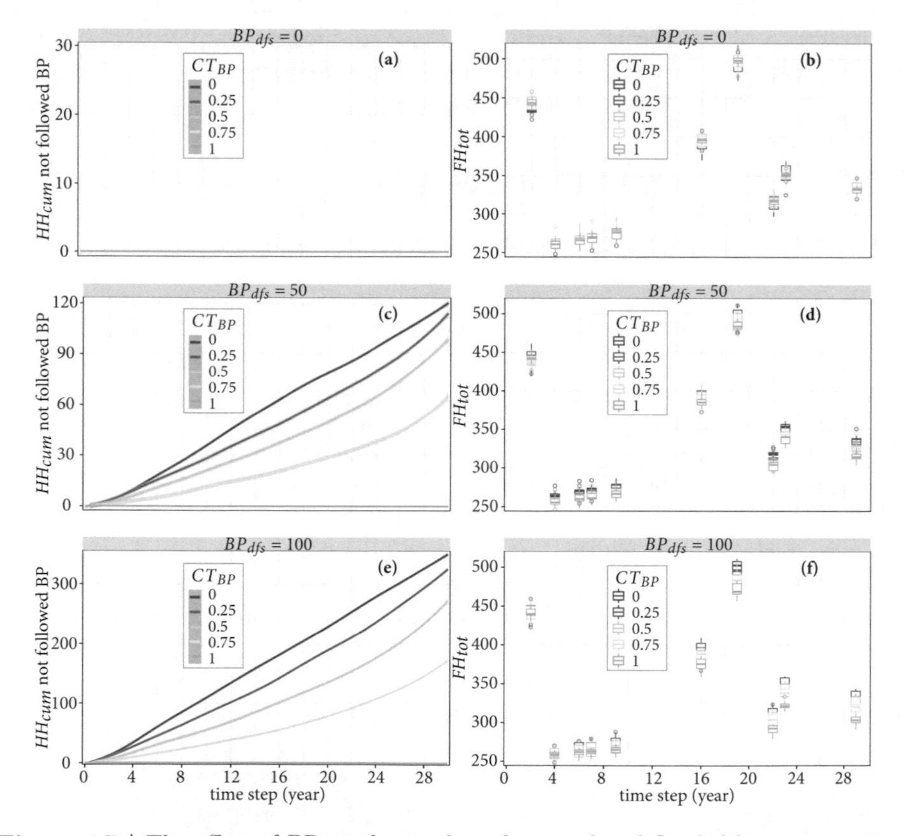

Figure 4.7 | The effect of BP on the number of exposed and flooded houses over time. All figures show BP compliance thresholds between 0 and 1. (a), (c) and (e) show the number of houses that do not follow the BP for BP_{dfs} values of 0 m, 50 m and 100 m, respectively, whereas (b), (d) and (e) show the total number of flooded houses for the same BP_{dfs} values. For these figures, the CT_{FZ} and CT_{BO} values are 0 and 0.5, respectively, and without structural measures.

The effect of the Flood Zones and Building Ordinance on the vulnerability of houses

The results in Figure 4.8(a) and (b) show the effect of the FZ and BO on the vulnerability of household agents, which is measured in terms of the number of (not-) elevated houses. The figures show a linear relationship between HH_{cum} that do not follow FZ and BO and the increase in the values of CT_{FZ} and CT_{BO}, respectively. The figures also illustrate that BO influences a larger number of agents than the FZ (for example, by more than 25 times at the end of the simulation for compliance threshold values of 0.5). It should be noted that Figure 4.8(b) does not include the initial number of houses that do not follow the BO, and it only shows the result after the simulation starts as in the case of non-compliance of the FZ in Figure 4.8(a).

Further, for the same CT_{FZ} and CT_{BO} values, not complying with the BO results in a higher number of flooded houses compared to not complying with the FZ. For example, Figure 4.8(c) and (d) show that for CT_{FZ} and CT_{BO} values of 0.5 (i.e., about 50 % compliance/enforcement), the median number of flooded houses that do not follow BO is about 30 times the number that do not follow FZ at *time step* = 29.

For both institutions, HH_{cum} and the number of flooded houses is higher with lower compliance thresholds (i.e., low policy compliance/enforcement). This is more important with bigger flood events and towards the end of the simulation as more vulnerable household agents are affected by the flood hazard. For example, as illustrated in Figure 4.9, with the increase in the CT_{BO} value from 0.5 to 1, the number of potentially vulnerable and flooded houses decreases. Regarding the effect of a change in compliance threshold values, not enforcing/complying with the BO results in more flooded houses than not enforcing/complying with the FZ. The main reason is that the BO applies to the whole island, affecting all agents while the FZ affects small portions of the island (see Figure 4.1 and Figure 4.8(a) and (b)).

The wider impact of complying with BO is again illustrated in Figure 4.8(e) and (f). Considering exposed houses (i.e., those houses that registered 5 cm or more flood), the median number of houses that are not flooded as they are elevated by 20 cm is about 20 times the number of agents that comply with FZ but not flooded for CT_{FZ} and CT_{BO} values of 0.5. However, Figure 4.8(g) and (h) show that for the same CT_{FZ} and CT_{BO} values, FH_{tot} are similar. This is because the effect of not enforcing/complying with the FZ on FH_{tot} is very small.

The figures also show that, regardless of the institution, there is an increase in FH_{tot} even when the compliance thresholds and the rainfall recurrence intervals are the same. For example, in Figure 4.8(f), the median FH_{tot} increases by about 27 % between *time step* = 4 and *time step* = 29, when CT_{BO} is 0.5 and the rainfall event in both time steps has a recurrence interval of 5-year. This is mainly attributed to the increase in the number of new houses in areas exposed to flooding.

The median FH_{tot} also increases by about 12 % even if the rainfall event is lower in intensity, as in the case of *time step* = 2 and *time step* = 19. Though the rainfall recurrence interval is reduced from 100-year to 50-year, FH_{tot} increases as the flood depth is high and the extent is large enough to affect more houses when the number of new houses increases.

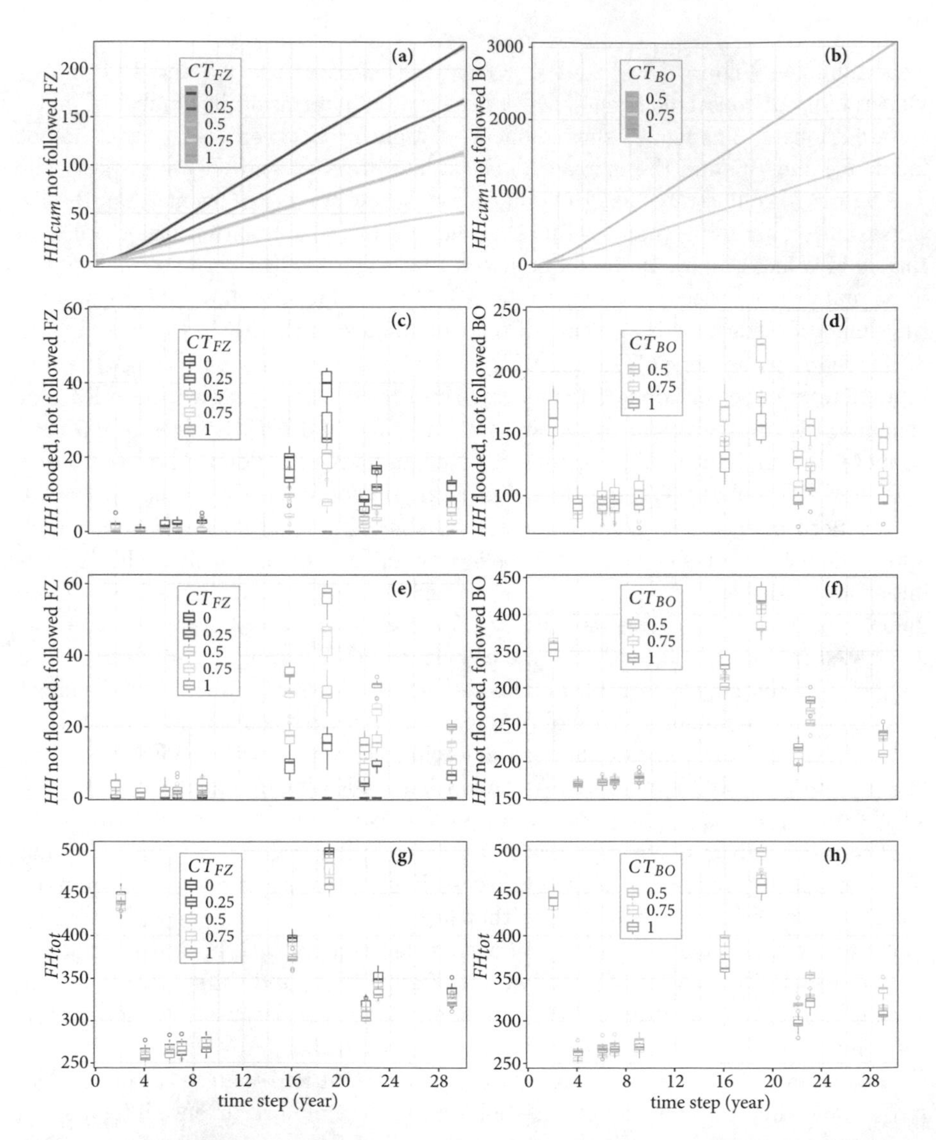

Figure 4.8 | The effects of FZ and BO on the number of vulnerable and flooded houses over time. (a) and (c) show the cumulative number of houses and number of flooded houses that do not follow FZ, respectively. (b) and (d) show similar results but when household agents do not follow BO. (b) does not include the initial condition. (e) and (f) show number of houses that followed FZ and BO, respectively, exposed in a flood event but not flooded as they are elevated. (g) and (h) show the total number of flooded houses for ranges of compliances of FZ and BO, respectively. For (a), (c), (e) and (g), CT_{BO} is 0 and for (b), (d), (f) and (h), CT_{FZ} is 0. For all the figures, BP_{dfs} is 50 m, CT_{BP} is 0 and no structural measures implemented.

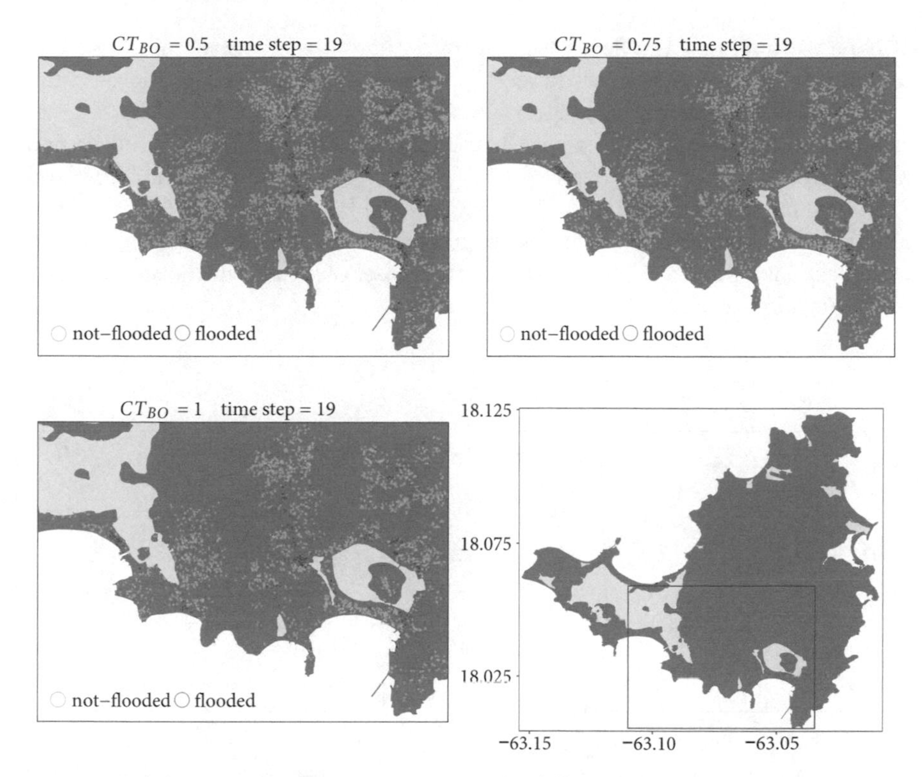

Figure 4.9 | Maps showing houses that do not follow the BO and (not-) flooded at *time step* = 19. The CT_{BO} values for (a), (b) and (c) are 0.5, 0.75 and 1, respectively. CT_{FZ} and CT_{BP} are 0 in the three cases. (d) shows part of Sint Maarten (red rectangle) plotted in (a), (b) and (c).

The effect of the structural measures on the hazard

The fourth institution tested is the implementation of structural measures. As shown in Figure 4.10, when flood hazard reduction measures are implemented, FH_{tot} decreases significantly compared to the results shown above. For example, comparing Figure 4.7(d) and Figure 4.10(a) or Figure 4.8(h) and Figure 4.10(b), FH_{tot} reduces by more than a half starting from *time step* = 7. In addition, comparing of *time step* = 2 and *time step* = 19, Figure 4.10(c) shows that with the implementation of structural measures, the number of flooded houses reduces. The reason is that the structural measures reduces the flood hazard (i.e., flood depth and extent), which in turn, also reduces the exposure of houses.

However, there are still flooded houses, especially along the coast, as shown in Figure 4.10(c). This is because the measures are not implemented in all catchments. For example, a coastal flood reduction dyke is implemented only in one catchment (see the difference between *timestep* = 4 and *time step* = 29 in Figure 4.10(c)), hence, other coastal areas register flooded houses.

Finally, Figure 4.11 shows the total number of houses on the island together with the elevated and the flooded houses over time in the "worst" and "best" case scenarios. The two scenarios are formed by taking the lowest and the highest values of the variable ranges in Table 4.4, respectively. The total number of houses in 30 years is lower in the best case as more households followed the BP and did not build houses. But, the number of elevated houses, complying with BO and FZ, is larger in the best case. As all household agents follow the BP, FZ and BO, and structural measures are implemented, the exposure, vulnerability and flood hazard are reduced in the best case. Hence, FH_{tot} is lower in that case, especially in the second half of the simulation period.

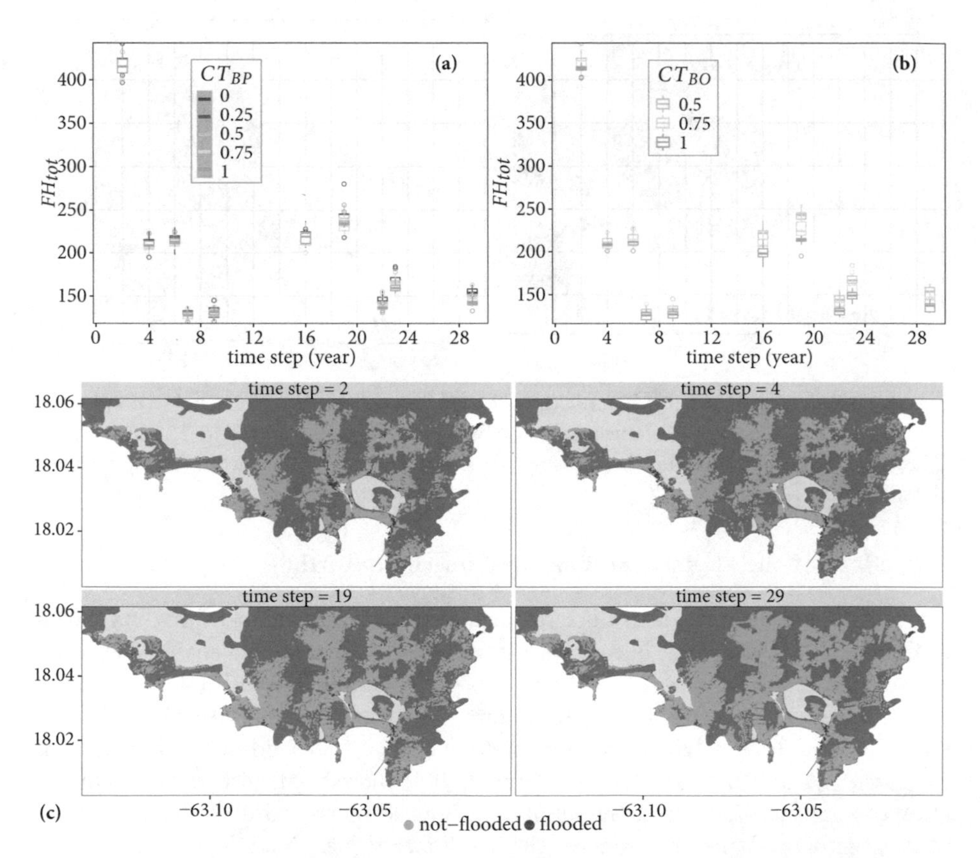

Figure 4.10 | The effect of structural measures on the number of flooded houses. (a) and (b) show the total number of flooded houses for ranges of compliances of BP and BO, respectively. For (a), CT_{BO} is 0.5 and for (b), CT_{BP} is 0. For all the figures, BP_{dfs} is 50 m, CT_{FZ} is 0 and structural measures are implemented. (c) shows maps of flooded and not-flooded houses at different time steps.

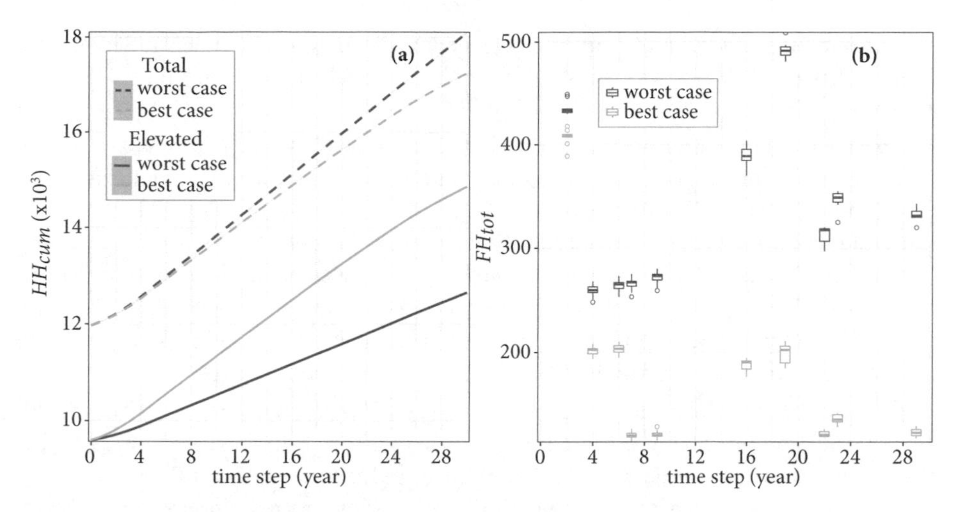

Figure 4.11 | (a) total and elevated number of houses (cumulative) and (b) total number of flooded houses in the "worst" and "best" simulation cases. In the "worst" case, the variable settings are: CT_{BO} is 0.5, CT_{BP} is 0 m, BP_{dfs} is 0, CT_{FZ} is 0 and no structural measures; whereas, in the "best" case, the variable settings are: CT_{BO} is 1, CT_{BP} is 1, BP_{dfs} is 100 m, CT_{FZ} is 1 and with structural measures.

4.5 Discussion and conclusion

The chapter presented a coupled ABM-flood model developed using the CLAIM framework to study FRM comprehensively. The coupled model examines existing and draft FRM policies in the Caribbean island of Sint Maarten. It also presented model evaluations in the form of uncertainty and sensitivity analysis, and model experimentations by defining policy enforcement/implementation scenarios. The four institutions considered in the model conceptualization are the Beach Policy, the Building and Housing Ordinance, the Flood Zoning and the hazard mitigation structural measures. In the experimentation and the analysis of the model, emphasis is given to degrees of compliance or enforcement of the institutions by heterogeneous household agents. The contribution of housing development to the flood risk is highlighted as well.

The conducted model evaluation shows that the coupled model output is affected by the flood model grid resolution. Fixing all other coupled model inputs, in general, the coarser the grid size, the lower the number of flooded houses. However, the use of a coarser grid significantly reduces the computation time. This is a relevant aspect, especially considering the need to replicate the coupled model due to the stochasticity of the ABM. Therefore, one should be careful when selecting the flood model grid size to balance the accuracy of model output and the total computation time.

Furthermore, the sensitivity analysis indicates that the coupled model output is sensitive to the initial number of households, the initial number of elevated houses and the increase in catchment imperviousness. Hence, collecting better quality datasets of existing houses, and acquiring better knowledge on how much a new house contributes to the imperviousness of a catchment will improve the model output analysis.

In general, simulation results show that when there is strict enforcement of the policies, which are manifested in higher compliance thresholds, communities' exposure and vulnerability reduces as more people follow the policies. That means, the number of potentially flooded houses decreases. This is observed mainly during bigger flood events (for example, at $time\ step = 19$) as their flood extents cover large area affecting a higher number of new household agents. However, in absolute terms, the significance of policy enforcement in reducing the flood risk depends on the aim and conditions of the institutions.

Because of the wider effect of the Building and Housing Ordinance, if household agents fully comply with the ordinance or if there is strict enforcement by VROMI, the ordinance has an important contribution in reducing the vulnerability of residents. Even when houses are exposed to flooding, they are not flooded as they are elevated. The number of exposed but not flooded houses (because they followed the ordinance) is slightly less than the total number of flooded houses. However, there are houses that are elevated but flooded in areas where the flood depth is greater than 0.2 m. This shows that the ordinance is not fully effective, although all agents comply with it.

On the other hand, with its localized effect, the Flood Zoning reduces the vulnerability of household agents located only in the delineated flood zones. The zones are already populated, and there is no much housing development in those areas. Hence, the policy's effect on reducing the total flood risk is low. In contrast to the Building and Housing Ordinance, the implementation of the Flood Zoning is beneficial within its area of effect as household agents will not likely flooded if they comply with it. That is because it obliges house floors to be elevated as high as 0.5 m to 1.5 m. However, it should also be noted that the policy is in draft stage and based on field observation and expert discussions, it would be challenging to convince developers to elevate building floors to such height as it is costly.

Similarly, the Beach Policy also has a localized effect, and its contribution to the overall flood reduction is low. Most parts of the Sint Maarten coast, especially where there are sandy beaches, are already occupied. As the properties developed in those coastal areas are of high value (most are hotels and service providers related to the tourism industry), a flooded property may result in bigger damage and loss. Hence, the policy can be an important institution if impacts are measured based on monetary values.

The simulation results indicate that the structural flood hazard reduction measures are the most important institution to reduce the flood risk. Upgrading channels cross-sections and building coastal flood reduction measures such as a dyke reduce the flood hazard, hence, reducing the number of flooded houses. Coastal measures are not often considered in Sint Maarten as there is a consensus that the measures may reduce the beauty of the beaches, hurting the tourism economy. However, as shown in the modelling scenarios, these measures are important to reduce the flood hazard unless all exposed buildings are demolished, and a strict policy that prohibits any construction along the coast is implemented.

Therefore, given the model setup and scenario simulations, implementing hazard reduction measures as well as strict enforcement of the Building and Housing Ordinance have a more substantial effect of reducing the number of flooded houses. However, the results and analysis of the coupled model outputs are subject to the challenges and limitations of modelling.

Models are abstractions of reality and they should not represent every feature of the system. Thus, assumptions are important elements of a model. In the coupled model we developed, we made assumptions to reduce the complexity of the models. We also made some assumptions merely because of lack of data. For example, had reliable data on the use of buildings in Sint Maarten been available, agent types such as businesses and public entities could have been represented in the model.

Agents' behaviour, such as their decision making is also simplified because of limited data availability. For example, the influence of agent interactions on the decision to follow a policy can be incorporated in the ABM based on household survey data. Regarding the flood model, we only consider storm surges as sources of coastal flooding. Including wave actions and climate change impacts such as sea level rise scenarios may intensify the coastal flood hazard affecting more houses. In such a case, the significance of the Beach Policy could be higher.

Another limitation is that housing development is exogenously imposed. The locations of the urban expansion are predefined based on a master plan. Including or coupling an urban growth model that simulates multiple scenarios of urban growth may give a better insight into how human dynamics contributes to flood risk in Sint Maarten. In addition, empirical validation of the model results was a challenge because of the exploration of non-existing scenarios and a lack of data. For example, floods are generated using synthetic rainfall event series in which a rainfall event occurs only once in a given year.

Hence, instead of focusing on reproducing a historical event, we emphasize the usefulness of the model by involving experts during the model conceptualization and parameter setting, and by consulting with them whether the results are realistic. We also analysed the model results based on 20 replications for each parameter combination. Estimating the experimental error variance using a statistical measure such as the coefficient of variation of the outputs indicates that more than 20 replications are needed to analyse the output better. However, we select the number of replications mainly based on the practical constraints in the computational resources.

5

The role of household adaptation measures to reduce vulnerability to flooding: The case of Hamburg, Germany[1]

5.1 Introduction

One of the goals of FRM is evaluation of strategies, policies, and measures to foster flood risk reduction and promote continuous improvement in flood preparedness and recovery practices (IPCC, 2014b). As flood risk is a function of flood hazard and communities' exposure and vulnerability, one way of reducing flood risk is by reducing the vulnerability at the household level. Focusing on the physical and economic aspects, measures to reduce vulnerability include elevating houses, retrofitting, dry or wet floodproofing, insurance and subsidies. These measures either prevent flooding or minimize the impact. While measures such as subsidies are offered by authorities or aid groups, the decision to implement most adaptation measures is made at the household level.

Household adaptation behaviour is affected by many factors such as flood risk perception, experience with flooding, socioeconomic and geographic factors, reliance on public protection, and competency to carry out adaptation measures (Bubeck *et al.*, 2012). The current literature mainly makes use of empirical research to draw insights on the role of household adaptation behaviour to reduce flood risk (for example, Botzen *et al.*, 2019; Grahn and Jaldell, 2019; Grothmann and Reusswig, 2006; Poussin *et al.*, 2014; Schlef *et al.*, 2018). Nevertheless, modelling efforts that bring behavioural and physical attributes together can further enrich these insights and add even more knowledge by incorporating the complex reality surrounding the

[1]This chapter is based on: Abebe, Y.A., Ghorbani, A., Nikolic, I., Manojlovic, N., Gruhn, A. and Vojinovic, Z. (2020). The role of household adaptation measures to reduce vulnerability to flooding: a coupled agent-based and flood modelling approach. *Hydrol. Earth Syst. Sci. Discuss. (Accepted for publication)*. DOI: https://doi.org/10.5194/hess-2020-272

human-flood interactions.

One of the research gaps in the current literature that present models to study household flood adaptation behaviour (for example, Erdlenbruch and Bonté, 2018; Haer *et al.*, 2016) is that flood events are not included in the simulation models. These studies define flood experience as an agent attribute that is set initially and stays the same throughout the simulations. A household that was not flooded in past events may get flooded in the future and may re-evaluate previous adaptation decisions, which in turn necessitates that flood events are included in the modelling. The second gap is that the effects of an economic incentive on the adaptation behaviour of individuals have not been addressed in the models. Such an analysis would provide an understanding of how much incentives contribute to flood risk reduction.

This study aims to enhance the current modelling practices of human-flood interaction to address the shortcomings of the current literature and draw new insights for FRM policy design. To achieve this aim, we build a coupled ABM-flood model, which comprehensively includes the human and the flood attributes in a holistic manner (Vojinovic, 2015). The coupled ABM-flood model builds on empirical and modelling insights in the literature: (i) by presenting an integrated simulation model instead of only ABMs, and (ii) by testing the effects of economic incentives and institutional configurations that have not yet been studied in the context of household flood adaptation behaviour. We use the protection motivation theory (PMT) (Rogers, 1983) to investigate household-level decision making to adopt mitigation measures against flood threats.

More specifically, this study extends two studies presented in (Birkholz, 2014) and (Abebe *et al.*, 2019b). Birkholz qualitatively explored PMT to study household flood preparedness behaviour in the German city of Hamburg. Birkholz collected information on local communities' flood risk perceptions and flood preparedness using semi-structured interviews. The current study uses the qualitative study as a base to conceptualize and further explore the household flood preparedness behaviour in Hamburg using an ABM. Abebe *et al.* (2019b) employ the coupled flood-agent-institution modelling (CLAIM) framework developed in (Abebe *et al.*, 2019a) to conceptualize the agent-flood interaction by decomposing the system into five components — agents, institutions, urban environment, physical processes and external factors. Their main focus was to study the implications of formal rules as institutions. In contrast, the current study mainly investigates the effect of informal institutions in the form of shared strategies applying the CLAIM framework. Additionally, the study examines individual strategies that affect households' adaptation behaviour.

The remainder of the chapter is structured as follows: Section 5.2 describes the study area. Section 5.3 provides a brief description of PMT and explains how it is conceptualized for the study area. Section 5.4 discussed how CLAIM is used to decompose the system, the ABM and flood model setups, model evaluations and experimental setups. Section 5.5 presents the results of the modelling exercises, followed by a discussion of the implications of the study findings and conclusions in Section 5.6.

5.2 Study area

We develop a coupled ABM-flood model that uses PMT as a tool to model households' flood vulnerability reduction behaviour for the FRM case of Wilhelmsburg, a quarter of Hamburg, Germany. The Wilhelmsburg quarter is built on a river island formed by the branching River Elbe, as shown in Figure 5.1. Most areas in Wilhelmsburg are just above sea level. Thus, flood defence ring of dykes and floodwalls protect the quarter. In 1962, a hurricane-induced storm surge (5.70 m above sea level) overtopped and breached the dykes, and more than 200 people lost their lives and properties were damaged due to coastal flooding in Wilhelmsburg (Munich RE, 2012). As a result, the authorities heightened and reinforced the coastal defence system. According to the Munich RE report, after 1962, eight storm surges of levels higher than 5.70 m occurred (most between 1990 and 1999), but none of the events caused any damage as coastal protection has been improved.

Those events reminded residents of the potential risks of coastal flooding, while, at the same time, increasing their reliance on the dyke protection system. The reliance on public protection is promoted by the authorities, who do not encourage the implementation of individual flood risk reduction measures referring to the strength

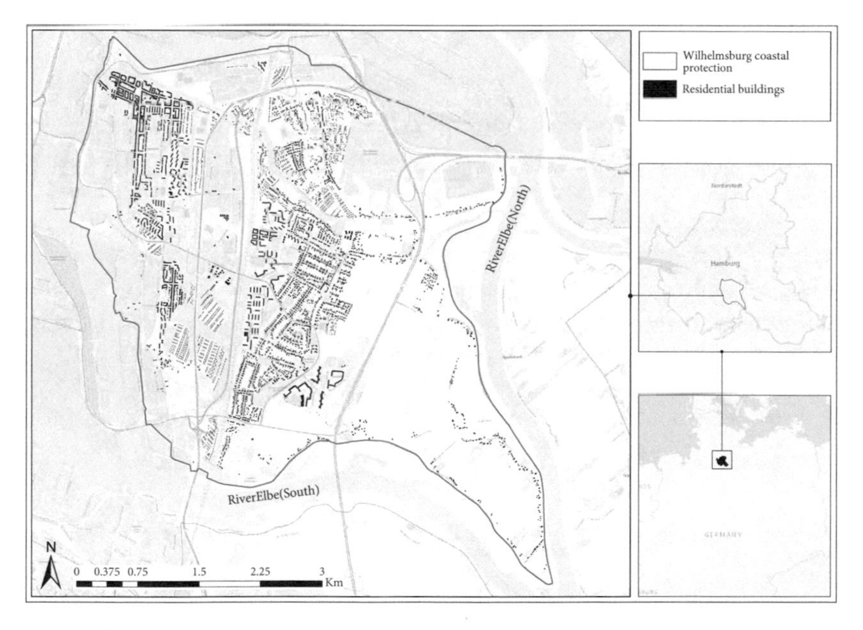

Figure 5.1 | A map of the study area of Wilhelmsburg. The red polygon shows Wilhelmsburg's coastal protection ring of dykes and walls. The study focuses on residential housings within the protected area. The buildings shown in the map are only those that are part of the model conceptualization. The inset maps in the right show the map of Germany (bottom) and Hamburg (top). (Source: the base map is an ESRI Topographic Map).

of the dyke system. On the other hand, the authorities disseminate warning and evacuation strategies to the public, acknowledging that there can be a flood in future. There is a probability that a storm surge bigger than the design period of the coastal defence may occur in the future, and climate change and sea level rise may even intensify the event. Hence, protecting houses from flooding should not necessarily be the responsibility of the authorities. Households should also have a protection motivation that leads to implementing measures to reduce flood risk.

5.3 Protection motivation theory

As shown in Figure 5.2, PMT has three parts — sources of information, cognitive mediating processes and coping modes (Rogers, 1983). The *sources of information* can be environmental such as seeing what happens to others and intrapersonal such as experience to a similar threat. Triggered by the information, the *cognitive mediation process* includes the threat and coping appraisals. The *threat appraisal* evaluates the severity of and the vulnerability to the threat against the intrinsic and extrinsic positive reinforcers. The *coping appraisal* evaluates the effectiveness of an adaptation measure to mitigate or reduce the risk, the ability to implement the measure, and the associated cost to implement the measure. If the threat and coping appraisals are high, households develop a *protection motivation* that leads to action. The *coping modes* can be a single act, repeated acts, multiple acts or repeated, multiple acts.

Originally developed in the health domain (Rogers, 1983), PMT has been extended and applied in diverse domains that involve a threat for which individuals can carry out an effective recommended response available (Floyd *et al.*, 2000). For example, in FRM studies, Poussin *et al.* (2014) extended the PMT by adding five factors — flood experience, risk attitudes, FRM policies, social networks and social norms, and socioeconomic factors — that directly determine the protection motivation of households. Two studies applied PMT in ABMs to test the effectiveness of flood risk communication strategies and the influence of social network on the adoption of protective measures to reduce households' vulnerability to flooding (Erdlenbruch and Bonté, 2018; Haer *et al.*, 2016). They compute the odds ratio and probability of implementation to model household decision on flood preparedness.

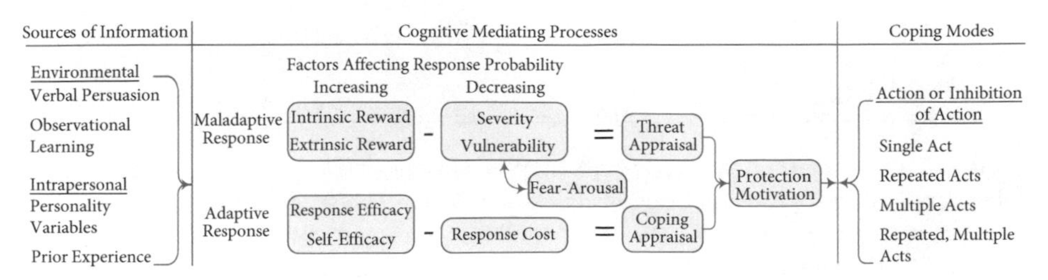

Figure 5.2 | The original schematization of the protection motivation theory (from Rogers, 1983)

One of the conclusions of the studies is that communication policies should have information regarding both the flood threat and coping methods to increase the adaptation rate.

Conceptualizing the protection motivation theory for Wilhelmsburg

In this study, we have modified the original PMT (Rogers, 1983) to use it in an FRM and ABM contexts for the specific case of Wilhelmsburg. In the original theory, the sources of information initiate both the threat appraisal and coping appraisal processes. However, in the current study, the sources of information influence the threat appraisal only. We assume that if there is a threat and need to implement a coping measure, the agents know the type of measure they implement based on their house categories (see Table 5.1).

In the threat appraisal, the *maladaptive response* is the current behaviour of not implementing household-level flood vulnerability reduction measures. In the case of Wilhelmsburg, the maladaptive response is affected by flood experience, reliance on public protection (i.e., the dyke system), climate change perception and source of information.

- The flood experience refers to any experience from which households can be directly affected by flooding, or they have witnessed flooding that affected others in Wilhelmsburg.

- The reliance on public protection is related to the flood experience. Residents of Wilhelmsburg who have not experienced flooding have a high reliance on the dyke system. The fact that seeing the dykes on a daily basis give residents a sense of protection and underestimate the flood threat. The reliance on public protection is also associated with the trust the residents have on the authorities when it comes to FRM. However, as some informants who experienced the 1962 flood described, the reliance on the dyke system drops if flooding occurs in the future (Birkholz, 2014).

- We include agents' climate change perception as a factor as some residents of Wilhelmsburg described that sea level rise might increase the occurrence of flooding in future. The effects of climate change create some discomfort and stress, and hence, it is seen as a source of concern. Besides, Germans, in general, are concerned about climate change in which 86 % are "extremely to somewhat worried" (NatCen Social Research, 2017).

- The source of information is an important factor that shapes residents perception of flood risk. The municipal and state authorities have a firm belief that the dyke system is the primary flood protection measure, and there is no need to implement individual measures to protect properties. However, these authorities communicate evacuation strategies in case the dykes fail or overtopped by a storm surge. On the other hand, other sources such as experts from the Technical University of Hamburg-Harburg organized flood risk awareness workshops presenting the flood risk in Wilhelmsburg and different adaptation measures that individuals could implement. Media also has a role in creating concern by showing flooding and its impacts in other German cities and even other countries.

In the coping appraisal, the *adaptive response* is developing a protection motivation behaviour to implement flood vulnerability reduction measure. The factors that affect the response probability in this conceptualization are personal flood experience, house ownership, household income, subsidy from the state and social network.

- Personal flood experience refers to a direct flood experience in which an agent's house was flooded before. It is a major factor that drives the adaptive response (Bubeck *et al.*, 2012). The factor is used as a proxy for behaviours in case of near-miss flood events as agents tend to make riskier decisions if they escape damage while others are flooded (Tonn and Guikema, 2017).

- We include house ownership as a factor though it has a small to medium effect on the adaptive response (Bubeck *et al.*, 2012). However, this factor is also used as a proxy for tenancy, which is an important factor since tenants tend not to implement measures. Hence, house ownership in this context specifies whether an owner or a tenant occupies a house at a given time.

- Household income has a significant influence on the adaptive response especially when agents implement measures that bring structural changes or adjustments to buildings such as flood proofing and installing utility systems to higher ground (Bubeck *et al.*, 2013). Hence, this factor affects only those households that intend to implement "structural measures".

- The subsidy is any financial help the authorities may provide to encourage implementation of individual adaptation measures. Currently, the authorities do not provide subsidies as they invest only on public protections. But, the assumption is that if a future low probability storm surge overtop or overflow the dyke system and flooding occurs, the authorities may take responsibility for the damages of properties given their assurance that people are safe and do not need to implement individual measures. As the subsidy is financial support, we conceptualize this factor similar to the household income affecting household agents that implement structural measures.

- The social network factor represents agents' relatives, friends or neighbours who have implemented any adaptation measure. Bubeck *et al.* (2013) showed that residents conform to the protection mitigation behaviour of others in their social network.

The state subsidy and the household income are proxy measures for the financial *response cost* of implementing the measures. In terms of other costs such as time and effort, we assume that the agents have no limitation. The assumptions related to response efficacy is that agents implement the adaptation measure specified in the shared strategy based on the type of houses they own, and the measure is assumed to be effective to reduce flood damage. However, it does not necessarily imply that the measure is the best possible. Similarly, the assumption related to self-efficacy is that either agents need to hire technicians that are capable of successfully implementing the measures or they are capable of implementing the measures by themselves. Appendix B lists the assumptions made to conceptualize and develop the model.

At last, *protection motivation* is an intention to implement coping responses (Rogers, 1983), which may not necessarily lead to actual behaviour (Grothmann and Reusswig, 2006). In our conceptualization, agents may delay the implementation of measures after they positively appraise coping. Agents may also change their behaviour through time and abandon temporary measures affecting their protection motivation.

5.4 Model setups

5.4.1 CLAIM decomposition

We use the CLAIM framework (Abebe *et al.*, 2019a) to decompose and structure the FRM case of Wilhelmsburg as CLAIM provides the means to explicitly conceptualize household behaviour and decision making, households interaction among themselves and with floods, and institutions that shape household behaviour. The primary source of the conceptualization is the doctoral dissertation by Birkholz (2014). Birkholz applied semi-structured, in-depth interviews with residents; academic and grey literature reviews; and personal observation of the study area. Besides, we use our local knowledge of the study area and performed formal and informal conversations with residents and authorities to develop a conceptual model.

1. Agents: The conceptual model includes two types of agents. The first ones are household agents that represent residents of Wilhelmsburg. The conceptualization focuses on residential buildings occupied by households. The model excludes businesses, industries, farmlands and other auxiliary buildings. The second agent is the representation of the Hamburg state authorities. This agent provides information about FRM. It may also assist household agents by providing subsidies so that they implement measures that reduce their vulnerability to flooding.

2. Institutions: In Wilhelmsburg, there is a common understanding that it is the responsibility of the authorities to protect the people. There is no institution, formal or informal, that influence household behaviour to reduce vulnerability. As a result, we will test hypothetical shared strategies that may have some effect on household agents flood risk. The conceptual model consists of five institutions in which one is related to the authority agent providing subsidies

to household agents and the rest related to households implementing vulnerability reduction measures depending on the house categories. The measures considered are installing utilities in higher storeys, flood adapted interior fittings, flood barriers and adapted furnishing.

3. Urban environment: The Wilhelmsburg quarter that is surrounded by the ring of dykes and walls defines the urban environment (see Figure 5.1). The household and authority agents live and interact in this environment. Coastal flooding also occurs in the same environment. In our conceptualization, we focus only on household behaviour to protect their houses. Therefore, the only physical artefact explicitly included in the conceptual model are residential houses. The adaptation measures that households may implement do not have physical representations in the model though their impact is implicitly evaluated if a house is exposed to flooding. Similarly, the dyke system is implicitly included in the hydrodynamic processes to set up the boundary conditions of overflow and overtopping discharge that causes coastal flooding. The conceptualization does not include any other infrastructure.

4. Hydrologic and hydrodynamic processes: Located in the Elbe estuary, the main physical hazard that poses a risk on Wilhelmsburg is storm surge from the North Sea. If the surge is high or strong enough to overtop, overflow or breach the dykes, a coastal flood occurs. The study only considers surge induced coastal flooding due to dyke overtopping and overflows.

5. External factors: The source of a flood is an extreme storm surge induced by wind, tide and cyclone. As there is no major risk of pluvial flooding, we exclude rainfall as a flood source. Regarding external political and economic factors, though there is a European Union Floods Directive that requires member states such as Germany to take measures to reduce flood risk, it does not specify the type of measure implemented. In Wilhelmsburg, the authorities invest primarily on the dyke system; hence the implications of the Floods Directive on individual adaptation measure is not relevant in this study. Furthermore, subsidies that may be provided by the government are included within state subsidies in our conceptualization.

5.4.2 Agent-based model setup

As described in Chapter 3, the MAIA (Modelling Agent systems using Institutional Analysis) meta-model (Ghorbani *et al.*, 2013) is used to structure the human subsystem. Below, we will describe the FRM case of Wilhelmsburg using the social, institutional, physical and operational structures of MAIA.

Social structure This structure defines agents and their states and behaviours. In the CLAIM decomposition, we defined two types of agents — the household and the authority agents.

- The household agents are representations of the residents of Wilhelmsburg. These agents live in houses. The actions they pursue include appraising threat and coping, implementing adaptation measure, and assessing direct damage. The agent attributes related to threat appraisal are flood experience, reliance on public protection, perception of climate change and source of information about flooding. The attributes related to coping appraisal are direct flood experience, house ownership and household income. If agents decide to implement an adaptation measure, they know which measure to implement based on the institutions identified.

- The authority agent represents the relevant municipal and state authorities that have the mandate to manage flood risk in Wilhelmsburg. This agent does not have a spatial representation in the ABM. The only action of this agent is to provide subsidies to household agents based on the policy lever defined in the experimental setup of the ABM. We model subsidy in a more abstract sense that if agents receive a subsidy, they implement an adaptation measure assuming that agents are satisfied with the amount they receive.

Physical structure This structure defines the physical artefacts of the system. In this study, the physical components that are modelled explicitly are residential houses and the urban environment. Houses spatially represent the household agents in the ABM. They have geographical location represented using polygon features, as illustrated in Figure 5.1. These polygons are used to compute the area of the houses.

Houses also have types, which are classified based on "the type of building, occupancy of the ground floor and the type of facing of the building" (Ujeyl and Rose, 2015, p. 1540006–6). This study includes 31 types of houses, which we group into five categories: single-family houses, bungalows, IBA buildings, garden houses, and apartment/high-rise buildings. Appendix C provides a complete list of the 31 types of houses. If a house is flooded, the potential building and contents damages of the house are computed in monetary terms based on the house type.

The urban environment is the area enclosed by the ring of dykes, as shown in Figure 5.1. A raster file represents the environment, and if floods occur, agents obtain information about flood depth at their house from the environment. The adaptation measures are also physical components of the system, but they are represented in the ABM only implicitly.

Institutional structure This structure defines the institutions that govern agents' behaviour. Institutions in CLAIM are coded using the ADICO grammar, which refers to the five elements institutional statements might contain: Attributes, Deontic, aIm, Condition and "Or else" (Crawford and Ostrom, 1995). Table 5.1 shows the five institutional statements that influence the implementation of individual flood risk reduction measures. When an agent is permitted to do an action (deontic *may*) with no explicit sanction (no "or else") for failing to do the action, the statement is referred to as a *norm*. In this case, the last institutional statement related to the subsidies is conceptualized as a norm. The authority agent may give subsidies, but it is not obliged to do so and faces no sanction if it decides not to provide subsidies. When the deontic and "or else" components are absent from an ADICO statement, the statement is referred to as a *shared strategy.* Therefore, the first four statements in Table 5.1 are shared strategies as there are no sanctions for non-compliance with the statements (no "or else" component), and there are no deontic. When a shared strategy drives a system, agents do what the majority in that system does. As a result, a household implements a measure when the majority of households implement the adaptation measure. However, the household also has the option not to implement the measure without incurring any punishment.

In our conceptualization, households implement a specific primary measure or a secondary measure (stated in the "aim") based on the category of a house they occupy (stated in the "condition"). Considering primary measures, as most single-family houses in Wilhelmsburg have two or three floors, household agents that live in such houses install utilities such as heating, energy, gas and water supply installations in higher floors. Household agents that live in bungalows and IBA buildings implement flood adapted interior fittings such as walls and floors made of water-

Table 5.1 | ADICO table of institutions for the Wilhelmsburg FRM case.

Attributes	Deontic	aIm	Conditions	Or else	Type
Households		install utilities in higher storeys	if they live in single-family houses		Shared strategy
Households		implement flood adapted interior fittings	if they live in bungalows and IBA buildings		Shared strategy
Households		implement flood barriers	if they live in garden houses, apartments and high-rise buildings		Shared strategy
Households		implement adapted furnishing as a secondary measure	if they have already implemented a measure and if they do not live in bungalows and garden houses		Shared strategy
Authority	may	provide subsidies to households to implement measures	e.g., if houses are flooded		Norm

proofed building materials. Agents that live in garden houses and apartment/high-rise buildings implement flood barriers. The barriers implemented by garden houses are sandbags and water-tight windows and door sealing while the latter implement flood protection walls. Household agents that have already implemented a primary measure may also implement a secondary measure. This measure is adapted furnishing, which includes moving furniture and electrical appliances to higher floors. As most bungalows and garden houses are single-storey housings, they do not implement adapted furnishing.

Installing utilities in higher floors and flood adapted interior fittings are permanent measures that alter the structure of the house, and we assume that once they are implemented, they will not be abandoned. Therefore, in PMT terminology, implementing these measures is a *single act coping mode.* In contrast, flood barriers and adapted furnishing are temporary measures in which agents must decide whether to implement them every time, just before a flood event. Therefore, implementing these measures is a *repeated acts coping mode.* Implementing both primary and secondary measures is a *repeated, multiple acts coping mode.*

Operational structure This structure defines the agents' actions and their interactions with other agents and the environment. The model implementation flow chart shown in Figure 5.3 lays out the actions agents perform at every time step. First, household agents assess if they perceive flood as a threat. If they do, they appraise coping that leads to protection motivation behaviour. Second, if there is the intention to implement a measure, they implement the adaptation measure specified in the institutional structure. Lastly, if there is a flood event at a given time step, the house layer is overlaid with a flood map corresponding to the event. Households check the flood depth at their property and assess the building and contents damages. Agents' attributes are updated if the actions change their states. This process is performed until the end of the simulation time. We will describe below how the actions — threat appraisal, coping appraisal, adaptation measure implementation, damages assessment and measures abandoning — are evaluated in the model.

Action 1: threat appraisal

In the ABM, the factors that affect household agents perception of flood threat in Wilhelmsburg are their flood experience (FE), their reliance on public protection (R), mainly the ring of dykes, their perception of future climate change (CC) and their source of information (SoI). Household agents update the four factors every time step based on the following criteria:

- FE is related to whether an agent lives in Wilhelmsburg when a flood event happens, and it has a binary value of `Yes` and `No`. The value of FE changes only after a flood event as given in Eq 5.1. Since the last major flood occurred in 1962 and only 14 % of Wilhelmsburg's residents are older than the age of

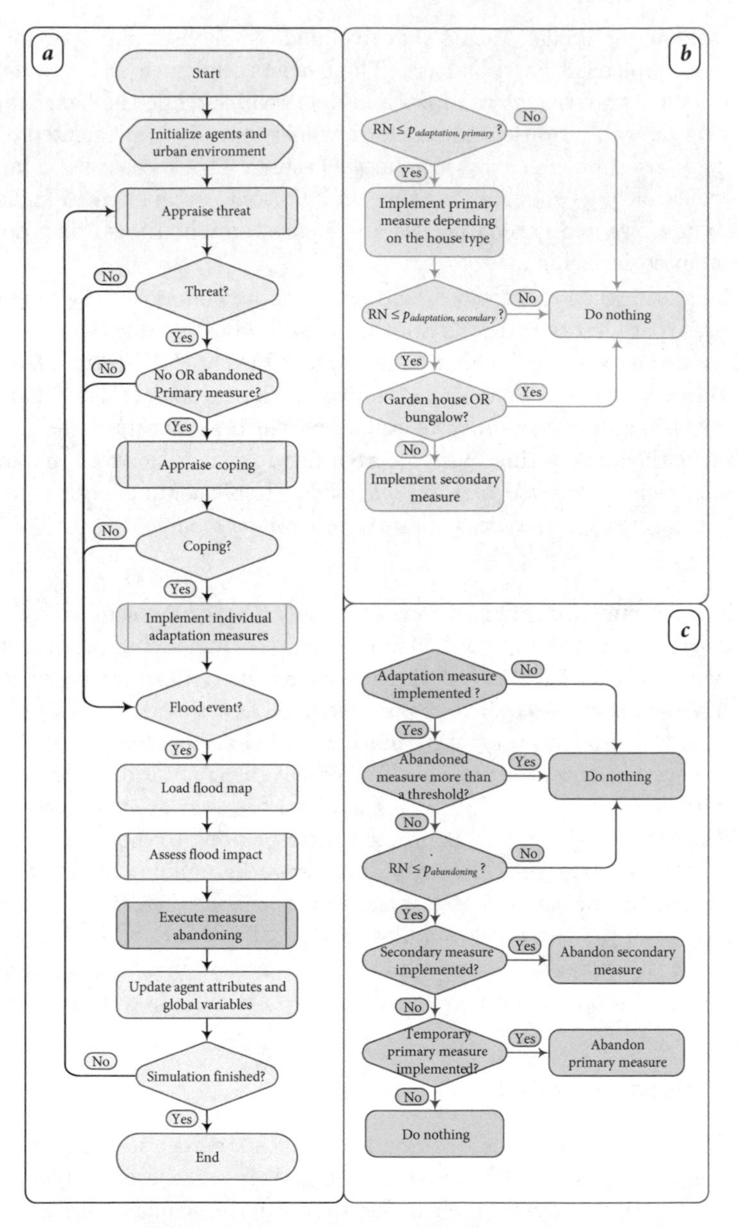

Figure 5.3 | CLAIM model implementation flowchart for the FRM case of Wilhelmsburg. (a) shows the general flow chart. (b) shows how implementing individual adaptation measures is modelled in the ABM while (c) shows how measures abandoning is modelled. The rest of the actions shown in sub-process shapes in (a) (shapes with double-struck vertical edges) are shown in figures below. In (b) and (c), RN is a random number, $p_{adaptation,primary}$ and $p_{adaptation,secondary}$ are the probabilities of adapting primary and secondary measures, respectively, and $p_{abandoning}$ is the probability of abandoning a primary and secondary measure.

65 (according to the 2011 census[2], the FE attribute of 86 % of the agents is randomly initialized as `No`. We assume that the flood experience does not fade over time.

$$FE = \begin{cases} Yes & \text{if agent lives in Wilhelmsburg when flood occurs} \\ No & \text{otherwise} \end{cases} \tag{5.1}$$

- R has a value of `Low`, `Medium` and `High`. It is dependent on FE and whether an agent has direct flood experience (see Eq 5.2). The `Medium` value reflects the uncertain position of agents towards the dyke system if they witness flooding in Wilhelmsburg. The value of R does not change unless there is a flood event and agents are flooded. This attribute is initialized based on the agents FE status.

$$R = \begin{cases} Low & \text{if } FE = Yes \text{ \& agent is flooded} \\ Medium & \text{if } FE = Yes \text{ \& agent is not flooded} \\ High & \text{if } FE = No \end{cases} \tag{5.2}$$

- CC has a value of `Yes`, `No` and `Uncertain`. The CC value of every agent is generated randomly from a uniform distribution, as shown in Eq 5.3. The thresholds in the equation are based on a study on country level concern about climate change in which 44 % Germans are "very or extremely worried", 42 % are "somewhat worried" and the remaining 14 % are "not at all or not very worried, or does not think climate change is happening" (NatCen Social Research, 2017). However, the study does not directly relate climate change with flooding. The value of this attribute may change over the simulation period. Assuming that agents may update their CC attribute at least once every Y_{CC} years, there is a probability of $1/Y_{CC}$ at every time step to update the attribute using Eq 5.3.

$$CC = \begin{cases} Yes & \text{if } Random \sim U(0,1) \leq 0.44 \\ Uncertain & \text{if } 0.44 < Random \sim U(0,1) \leq 0.86 \\ No & \text{if } Random \sim U(0,1) > 0.86 \end{cases} \tag{5.3}$$

- We broadly categorize SoI as `information from Authorities`, which informs agents that the dykes will protect everyone and there is no flood threat, and `information from other sources`, which informs agents that there can be a flood threat and agents need to prepare. SoI is assigned to agents randomly. Similar to the CC attribute, there is a probability of $1/Y_{SoI}$ to update the SoI attribute assuming that agents may update this attribute at least once every Y_{SoI} years.

[2]Interactive maps for Hamburg for the 2011 Census can be found at https://www.statistik-nord.de/fileadmin/maps/zensus2011_hh/index.html)

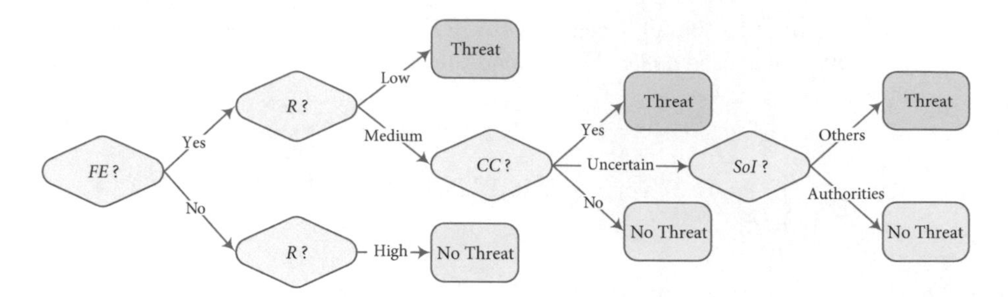

Figure 5.4 | Decision tree for the threat appraisal.

The flood threat is a function of the four factors and agents assess their perception of flooding as a threat using a rule-based decision tree (see Figure 5.4). If an agent has no experience of flooding, its reliance on public protection is high, and it perceives no threat of flooding regardless of the other factors. On the other hand, if an agent has low reliance on the dyke system, it perceives flooding as a threat regardless of the other factors. In case an agent's reliance on public protection is intermediate, its perception of climate change determines the threat appraisal. A concern regarding future impacts of climate change results in a perception of flood threat while no concern leads to no perception of the flood threat. If an agent is uncertain about climate change impacts, its source of information determines the threat appraisal. As some of the attributes of agents may change over time, all agents appraise threat at every time step.

Action 2: coping appraisal

Coping behaviour is initiated depending on agents' belief in their ability to implement a measure, agents' expectation that the measure removes the threat or improves the situation, and the perceived costs of implementation. In our model, coping appraisal is influenced by agents direct flood experience, i.e., if they had personal flood experience (PFE), house ownership (HO), household income (HI), state/government subsidy (SS) and the number of measures within agent's social network (SN).

- PFE has a value of `Yes` or `No` based on agents direct flood experience. This attribute is initialized as `No` for all agents. The value of PFE changes only when an agent's house is flooded after an event as given in Eq 5.4.

$$PFE = \begin{cases} Yes & \text{if agent has direct flood experience} \\ No & \text{otherwise} \end{cases} \tag{5.4}$$

- HO has a value of `Own` or `Rented`. According to the 2011 census, in Wilhelmsburg, the share of apartments occupied by the owners was 15 % while apartments rented for a residential purpose were 82 %. The remaining 3 %

were vacant. Based on that, in the ABM model, we randomly initialize 15 % of the households as owners of the houses they occupy while the remaining 85 % as renters, assuming that the 3 % vacant apartments can potentially be rented. In this conceptualization, we assume that the house ownership of a percentage of the household agents changes randomly, at every time step.

- HI has a value of `Low` or `High`. Since income is considered sensitive information, the data is not readily available. Hence, we randomly initialize 30 % of the agents as low-income households and the rest as high-income. Similar to the house ownership, we assume that the income of a percentage of the household agents changes randomly, at every time step. It should be noted that this factor affects the agents that implement permanent adaptation measures of installing utilities in higher storeys and flood adapted interior fittings, which are classified as structural measures (see Bubeck *et al.*, 2013, p. 1330).

- SS has a value of `Yes` or `No`. This variable is related to the last institution mentioned in Table 5.1. In the ABM setup, it is used as a policy lever to test the effect of subsidies on the implementation of structural adaptation measures.

- SN has a value of `Low` or `High`. As shown in Eq 5.5, this factor depends on the number of agents that implement a specific type of adaptation measure for a given house category. If the number is greater than a threshold, agents who occupy that same house category will have `High` SN value. Otherwise, SN is `Low`.

$$SN = \begin{cases} High & \text{if } NA_{measureType} \geq threshold \\ Low & \text{otherwise} \end{cases} \tag{5.5}$$

where, $NA_{measureType}$ is the number of agents that implement a specific type of measure depending on the category of a house they occupy.

Coping is a function of the five factors, and agents appraise their coping using a rule-based decision tree illustrated in Figure 5.5. For households that implement a structural measure, the full decision tree is evaluated while for those that implement temporary measures, shapes and lines in the dashed line are not assessed. If household agents have direct flood experience, the conditions that they would not intend to cope and implement a structural measure are if they occupy a rented house and (i) they have high income but have low SN, or (ii) they have low income and received no subsidy, or (iii) they have low income and received a subsidy, but have low SN. If agents live in their own house, the only condition that they would not intend to cope is if they have low income, received no subsidy and have low SN. In all the other cases, agents coping appraisal results in intention to cope. If agents do not have direct flood experience, the only case that they develop a coping behaviour is when the agents own the house they occupy and (i) they have high income and

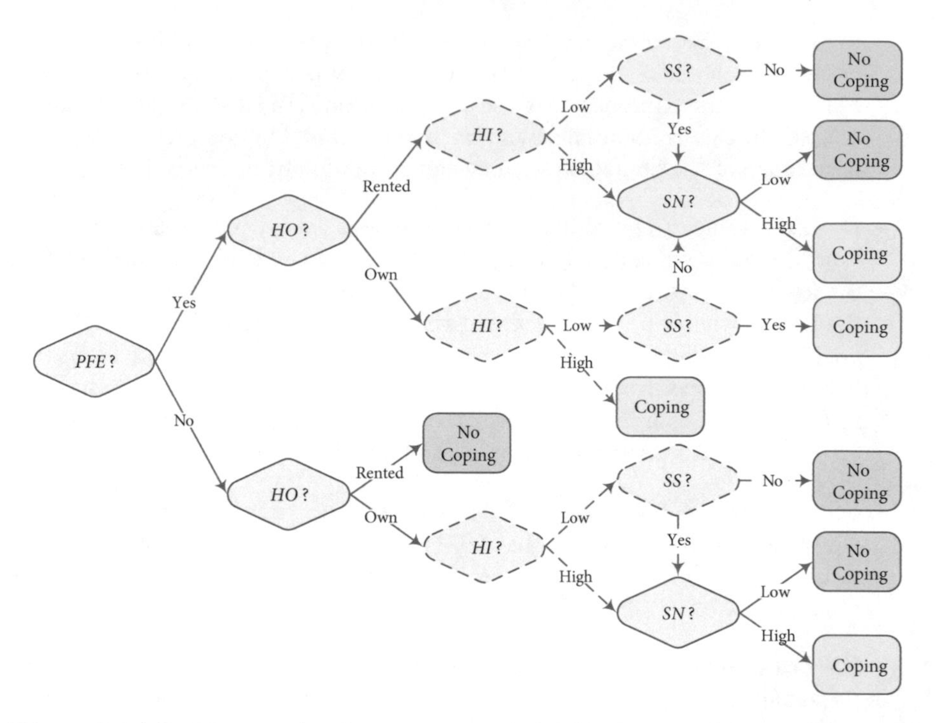

Figure 5.5 | Decision tree for the coping appraisal. The shapes and lines in dashed line are related to the income and subsidy factors, and they are executed only when households implement structural measures.

have high SN, or (ii) they have low income, have received a subsidy and have high SN. In the rest of the cases, household agents do not develop coping behaviour.

In the case of household agents that implement temporary measures, if the agents have direct flood experience, the only condition that they would not intend to cope is if they occupy a rented house and have low SN. If agents do not have direct flood experience, the conditions that they would not intend to cope is: (i) if they occupy a rented house and (ii) if they own the house but have low SN. In the rest of the cases, household agents develop coping behaviour.

An important aspect regarding the SN factor in our conceptualization is that its value is the same for all households who live in houses of the same category. That means, for example, if the value of SN is `High` for a certain house category, all households who occupy houses of that category will follow the same behaviour. But, as shared strategies drive the system in this case, households have the option not to develop that behaviour though most follow the crowd. To reflect this property of shared strategies, we introduce a shared strategy parameter (SSP) that works in tandem with the SN. The SSP is a kind of threshold that defines the percentage of household agents that follow the shared strategy. For example, if agents SN factor is `High`, they develop a coping behaviour when a randomly drawn number from a uniform distribution is less than or equal to a predefined value of SSP.

Action 3: household adaptation measure implementation

Following Erdlenbruch and Bonté (2018), we introduce a *delay parameter* that affects measures implementation. The delay parameter represents the average number of years agents take to transform a protection motivation behaviour into an action, which is implementing a primary measure. The probability that a motivated individual will adapt at a given year is computed as $p_{adaptation,primary} = 1/\textit{delay parameter}$. We also introduce a *secondary measure parameter* that determines whether agents implement secondary measures. This parameter is set as a threshold value defined by the modeller's estimation. As shown in Figure 5.3(b), agents consider implementing secondary measures only if they implement primary measures. The assumption is that those agents have already appraised coping positively and they may have a protection motivation to implement a secondary measure. As stated earlier, only multi-storey house categories implement secondary measures.

Action 4: damages assessment

The impacts of a flood event can be estimated by the direct and indirect damages of flooding on tangible and intangible assets. In this study, we measure the flood impact based on the potential direct damages which are caused by the physical contact of floodwater with residential houses. We estimate the building and contents damages using depth-damage curves developed for the 31 types of houses in Wilhelmsburg, as discussed in (Ujeyl and Rose, 2015). The building damages are related to replacement and clean-up costs, whereas the contents damages are related to replacement costs of fixed and dismountable furnishing. Figure 5.6 shows the depth-damage curves for the different house types.

If household agents implement adaptation measures, the building and contents damages of their house reduce. Based on empirical researches (Kreibich and Thieken, 2009; Poussin *et al.*, 2015), we compute the damages reduced as a percentage reduction of the ones presented in Figure 5.6. Installing utilities in higher storeys reduces the building damage by 36 while it has no impact on the contents damage reduction. Implementing flood adapted interior fittings reduces both damage by 53 %. Implementing adapted furnishing reduces the contents damage by 77 % while it has no impact on the building damage reduction. In the case of flood barriers, implementing sandbags and water-tight windows and door sealing reduces only the building damage by 29 % whereas implementing flood protection walls reduces the flood depth by a maximum of one meter.

Action 5: measures abandoning

We also introduce an *adaptation duration parameter* factor that affects measures abandoning, following Erdlenbruch and Bonté (2018). The adaptation duration parameter represents the average number of consecutive years a household agent implements an adaptation measure. It is used to estimate the probability that an agent abandons the measure at a given year. The likelihood that a motivated individual abandons a measure at a given year is computed as $p_{abandoning} =$

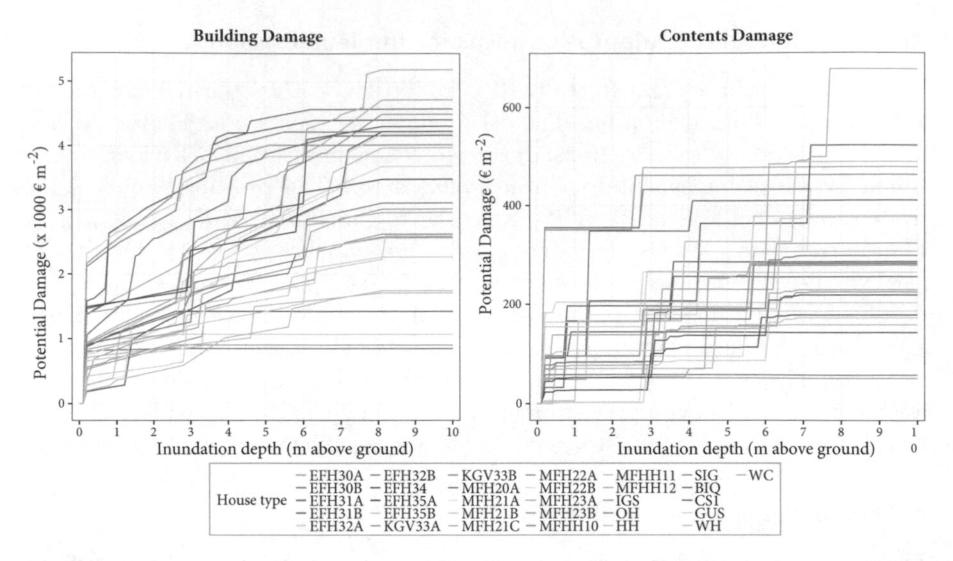

Figure 5.6 | Depth-damage curves for building (left panel) and contents (right panel) of 31 house types in Wilhelmsburg. A description of the house type codes is given in Appendix B.

1/*adaptation duration parameter*. This parameter affects only agents that implement temporary measures. The minimum adaptation duration would be one year. As shown in Figure 5.3(c), we limit the frequency of abandoning a measure by an agent using the *abandoning frequency threshold*. The assumption is that agents will not abandon a measure any more if they abandon and implement it a certain number of times specified in the threshold. If an agent has implemented a secondary measure, the first option to abandon is that measure. Otherwise, the agent abandons the temporary primary measure. In the latter case, the agent appraises coping once again.

Once the conceptual model is developed, we convert it to a programmed model using the Java-based Repast Simphony modelling environment (North *et al.*, 2013). The ABM software developed in this study, together with the ODD protocol (Grimm *et al.*, 2010) that describe the model, is available at https://github.com/yaredo77/Coupled_ABM-Flood_Model_Hamburg. The simulation period of the ABM is 50 time steps in which each time step represents a year. The number of household agents is 7859.

5.4.3 Flood model setup

The flood model in this study is based on extreme storm surge scenarios and 2D hydrodynamic models explained in (Naulin *et al.*, 2012; Ujeyl and Rose, 2015). The storm surge is composed of wind surge, local tides and a possible external surge due to cyclones. The extreme storm surge events are computed by considering the highest observed occurrence of each component. The three storm surge events — Event A, Event B and Event C — used in this study has a peak water level of 8.00 m, 7.25 m and 8.64 m, respectively (Naulin *et al.*, 2012). Numerical 2D hydrodynamic models are used to calculate water levels and wave stages around the dyke ring. In turn, these data are used to compute the overflow and wave overtopping discharges for the three scenarios.

To assess the flood hazard from the three scenario events, flood models that simulate coastal flooding are implemented. The model is developed using the MIKE21 unstructured grid modelling software (DHI, 2017b). The 2D model domain defines the computational mesh and bathymetry, in which the latter is based on a digital terrain model (see Figure 5.7). The surface resistance is expressed using a space-dependent Manning number that is based on the current land use categories. The time-dependent overflow and overtopping discharges over the dykes described above are used as boundary conditions. The output of the hydrodynamic model relevant for the current study is the inundation map showing the maximum flood depth in Wilhelmsburg. This is because the main factor that significantly contributes to building and contents damage is the flood depth (Kreibich and Thieken, 2009). Further, as houses are represented by polygon features (see Figure 5.1), the flood depth for a specific house is the maximum of the depths extracted for each vertex of the polygon that defines the house.

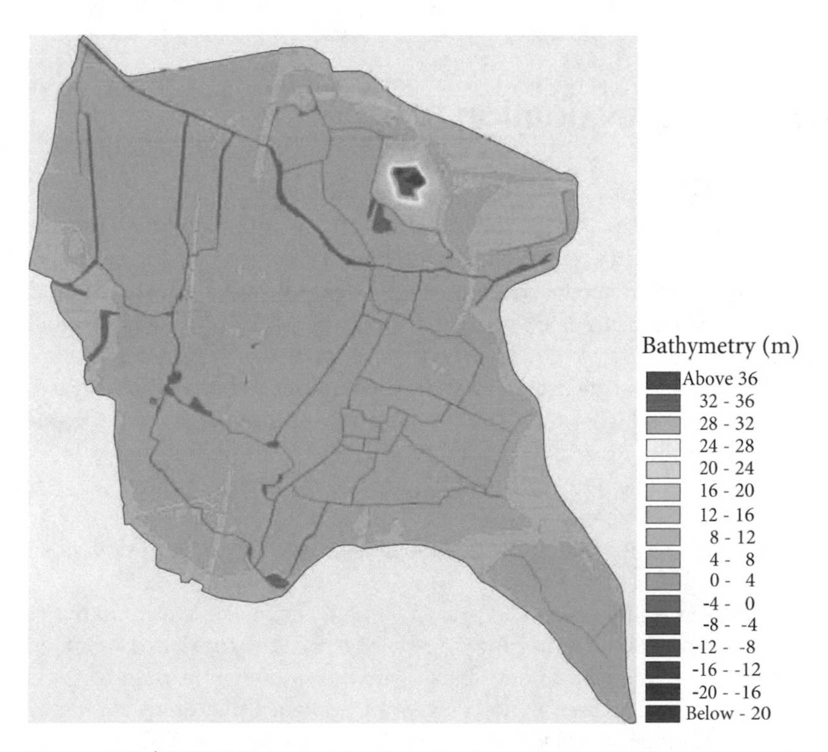

Figure 5.7 | MIKE21 coastal flood model domain showing the bathymetry.

5.4.4 Coupled model factors and setup

The input factors of the coupled ABM-flood model are presented in Table 5.2. The input factors are grouped into two. The first group includes the initial conditions and parameters that are regarded as control variables. Varying these factors is not of interest for the study; and hence, they are not included in the model experimentation. The second group comprises factors that are used to set up model experimentation and to evaluate the effect of household adaptation measures in FRM. In this group, the first three factors are related to the shared strategies defined in Table 5.1, while the last three are related to individual strategies. The flood event scenario is a randomly generated storm surge events series (see Figure 5.8). The percentage base values in Table 5.2 are respective to the total number of agents.

The response factors we use to measure the model outcome are the cumulative number of household agents that positively appraised coping ($Coping_{Yes}$), that positively appraised coping due to the social network element ($Coping_{Yes,SN}$), that implemented primary measures ($PM_{implemented}$), that abandoned primary measures ($PM_{abandoned}$), that implemented secondary measures ($SM_{implemented}$) and that abandoned secondary measures ($SM_{abandoned}$). In terms of damage, we focus on the building and contents damage mitigated rather than the total damage to highlight the benefits of household adaptation measures.

All simulations in this study are performed using the SURFsara high performance computing cloud facility (https://userinfo.surfsara.nl/systems/hpc-cloud).

5.4.5 Model evaluation

Model verification and validation

As mentioned in Section 5.4.3, the flood model we utilize in this study was developed and reported in a previous publication. Hence, we take the calibration and validation of the flood model at face value. Regarding the ABM, we model carried out verification using the evaluative structure of MAIA, which evaluates the relationship between agents' actions and expected response factors. For example, when agents implement measures, system-level number of secondary measures implemented cannot be higher than the primary measures implemented. Or, in coping appraisal, with an increase in the number of agents with high income, we expect a system-level rise in the number of coping agents. However, the average number of agents that implement permanent measures should not be influenced as there is no relationship between income and permanent measures implementation as specified in the conceptual model.

Validating the ABM model has proven to be a challenge. We validated the conceptual model using expert and local knowledge of the study area. But, we have not performed a statistical or data-based validation of the outputs of the simulation model. Currently, there is no practice of implementing household adaptation measures in Wilhelmsburg; thus, there is no data to perform such detailed model validation. Given the limitations, the practical purpose of the ABM is to showcase the

Table 5.2 | List of model input factors and their base values.

	Model input factors	Symbol	Base values[a]	Remark
Initial conditions and parameters	Initial percentage of households with FE	$FEthreshold_{ini}$	14 %	Based on 2011 census data (age group) and the last major flood in Wilhelmsburg
	Initial percentage of households with CC `Yes`	$CCthreshold1_{ini}$	44 %[b]	Based on NatCen Social Research, 2017
	Initial percentage of households with CC `Uncertain`	$CCthreshold2_{ini}$	42 %[b]	Based on NatCen Social Research, 2017
	CC update interval (years)	Y_{CC}	3	Authors estimation[d]
	SoI	SoI_{ini}	80 %	Authors estimation[d]
	SoI update interval (years)	Y_{SoI}	5	Authors estimation[d]
	Initial percentage of HO `Own`	HO_{ini}	15 %	Based on 2011 census data (apartments according to use)
	House ownership update	HO_{update}	1 %	Authors estimation[d]
	Initial HI `Low`	HI_{ini}	30 %	Authors estimation[d]
	Household income update	HI_{update}	1 %	Authors estimation[d]
	Abandon frequency threshold	$f_{abandoning}$	2	Authors estimation[d]
Factors for setting up model experiment	State subsidy	SS_{lever}	1[c]	Authors estimation[d]
	Shared strategy parameter	SSP	80 %	Authors estimation[d]
	SN threshold	$SN_{threshold}$	30 %	Authors estimation[d]
	Flood event scenario	$FE_{scenario}$	Scenario 1	Authors estimation[d]
	Delay parameter (years)	Y_{delay}	1	Authors estimation[d]
	Adaptation duration (years)	$Y_{adaptation}$	7	Authors estimation[d]
	Secondary measure parameter	SMP	30 %	Authors estimation[d]

[a] The percentage base values are respective to the total number of agents.

[b] The sum of the two CC thresholds should not exceed 100 %. If the sum is less than 100 %, the remaining is the percentage of agents who do not perceive CC as a source of threat.

[c] $SS_{lever} = 1$ refers to no subsidy.

[d] These estimations are based on authors expertise and knowledge of the study area.

benefits of household adaptation measures so that authorities and communities in Wilhelmsburg may consider implementing such measures to mitigate potential damages. Moreover, the model serves the purpose of advancing scientific understanding of socio-hydrologic systems, particularly human-flood interactions.

Estimating simulations repetition

ABMs are often stochastic For example, agent behaviours are determined based on random values generated from pseudo-random numbers, which produces results that show variability even for the same input factor setting (Bruch and Atwell, 2015; Lorscheid *et al.*, 2012; Nikolic *et al.*, 2013, p. 110–111). Hence, reliable ABM outputs are obtained by running simulations multiple times. To determine the number of simulation runs, we apply the experimental error variance analysis suggested by Lorscheid *et al.* (2012). The coefficient of variation (c_v) is used to measure the variability in the model output. Starting from a relatively low number of runs, the c_v of the model output is calculated by increasing the number of runs iteratively for the same factor settings. The number of runs is fixed when the c_v stabilizes or the difference between the c_v's of iterations falls below a criterion. This experiment is done for selected input factor settings to cross check whether output variations stabilize around the same number of runs irrespective of the factor settings. We evaluate the c_v's for the six response factors.

Sensitivity analysis

As in any model, the ABM developed in this study is subject to uncertainties. Regarding input factors uncertainty, the initial conditions and parameters mentioned in Table 5.2 are either based on our expert estimations or based on available coarse datasets such as the 2011 national census in Germany. Hence, an SA is carried out to allocate the model output uncertainty to the model input uncertainty. The SA method adopted in this study is the *elementary effects (EE) method*, also called the *Morris method* (Morris, 1991). The method is effective in identifying the important input factors with a relatively small number of sample points (Saltelli *et al.*, 2008, p. 109). Saltelli *et al.* explained that "the method is convenient when the number of factors is large [and] the model execution time is such that the computational cost of more sophisticated techniques is excessive" (p. 127). We employ this method because of the high computational cost related to the large number of simulation repetitions estimated (see Section 5.5.1).

The EE method is a specialized OFAT SA design that removes the dependence on a single sample point by introducing ranges of variations for the inputs and averaging local measures. The sensitivity measures proposed by Morris are the mean (μ) and standard deviation (σ) of the set of EEs, which are incremental ratios, of each input factor. In a revised Morris method, Campolongo *et al.* (2007) proposed an additional sensitivity measure, μ^*, which is the estimate of the mean of the distribution of the absolute values of the EEs. The sampling strategy to estimate the sensitivity measures is building r EE trajectories of $(k+1)$ points for each k factor, resulting in

Table 5.3 | Input factors considered in the sensitivity analysis, their distributions and value ranges. Factors specified in percentages are converted to decimals.

SA factors	Distribution	Range
$FEthreshold_{ini}$	Uniform	[0, 0.3]
$CCthreshold_{ini}$	Discrete	[1, 4][a]
Y_{CC}	Discrete	[2, 8]
SoI_{ini}	Uniform	[0.5, 1]
Y_{SoI}	Discrete	[3, 6]
HO_{ini}	Uniform	[0.1, 0.5]
HO_{update}	Uniform	[0, 0.02]
HI_{ini}	Uniform	[0.1, 0.5]
HI_{update}	Uniform	[0, 0.02]
$f_{abandoning}$	Discrete	[1, 4]

[a] The $CCthreshold_{ini}$ values for `Yes` and `Uncertain` are 0.35, 0.4, 0.45 and 0.5 for the discrete values of 1, 2, 3 and 4 respectively.

a total of $r(k+1)$ sample points. Following Saltelli *et al.* (2008, p. 119), we choose r to be 10, and each model input is divided into four levels within the input value range. In this study, the input factors selected for the SA are the initial conditions and parameters (as specified in Table 5.2). Therefore, the computational cost of the SA is $10(10+1) = 110$. In Table 5.3, we list these factors, their distributions and value ranges. In the SA, the other input factors presented in Table 5.2 are set to their base values.

5.4.6 Experimental setup

To evaluate the effect of the shared strategies listed in Table 5.2 and individual strategies such as delaying the implementation of measures, implementing secondary measures and abandoning measures, we set up simulations by varying the values of selected input factors as presented in Table 5.4. The subsidy levers 1, 2 and 3 represent no subsidy, subsidy only for flooded household agents and subsidy for all agents that consider flood as a threat, respectively. Considering the computational cost of simulations, we evaluate six flood event scenarios. The event series of the scenarios are randomly generated and shown in Figure 5.8. In these batch of simulations, all the other input factors are set to their base values, as stated in Table 5.2.

Table 5.4 | Input factors for model experimentation and their value ranges. Values of some factors are converted from percentages to decimals.

Symbol	Range	Step
SS_{lever}	[1, 3]	1
SSP	[0.5, 1]	0.1
$SN_{threshold}$	[0.2, 0.5]	0.1
$FE_{scenario}$	[1, 6]	1
Y_{delay}	[1, 10]	2
$Y_{adaptation}$	[3, 11]	2
SMP	[0, 0.6]	0.2

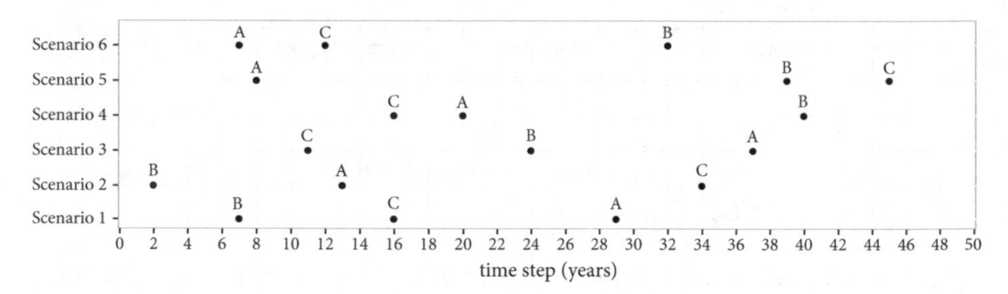

Figure 5.8 | Scenarios of flood events series. A, B and C represent flood events of storm surge with peak water levels of 8.00 m, 7.25 m and 8.64 m, respectively.

5.5 Results

5.5.1 Simulation repetitions

We iteratively run simulations starting from 100 to 5000 and compute the c_v's of six response factors for each iteration, for several input factor settings. As an example, Table 5.5 shows the c_v's for the factor setting in which all the input factors have the base values. Selecting a difference criteria of 0.001, the minimum sample size in which the c_v's start to stabilize is 3000. As the c_v's do not change while increasing the number of runs, we fix the number of runs to be 3000. For the following analysis (SA and policy-related experiments), simulation outputs are computed as averages of 3000 simulations per input factor setting.

Table 5.5 | Coefficient of variations (c_v) of response factors per iterations. The grey shaded area shows the number of runs in which the c_v's of all the response factors are stable for a difference criteria of 0.001.

	c_v per number of runs							
Response factors	**100**	**500**	**1000**	**1500**	**2000**	**3000**	**4000**	**5000**
$Coping_{Yes}$	0.015	0.034	0.029	0.027	0.027	0.027	0.027	0.027
$Coping_{Yes,SN}$	0.024	0.052	0.045	0.041	0.041	0.043	0.042	0.043
$PM_{implemented}$	0.015	0.034	0.029	0.027	0.027	0.027	0.027	0.027
$PM_{abandoned}$	0.163	0.171	0.170	0.170	0.169	0.169	0.169	0.168
$SM_{implemented}$	0.066	0.073	0.066	0.065	0.066	0.067	0.067	0.067
$SM_{abandoned}$	0.230	0.226	0.222	0.225	0.226	0.226	0.227	0.227

5.5.2 Sensitivity analysis results

The SA is carried out on 10 input factors, and the outputs quantify five response factors evaluated at *time step* = 50. Figure 5.9 shows the Morris sensitivity measures μ^* and σ plotted against each other for five response factors. As $Y_{delay} = 1$ in all the simulations, the response factors $Coping_{Yes}$ and $PM_{implemented}$ have exactly the same value. Hence, only the former response factor is displayed in the figure. The results show that the most important factor by far is HO_{update} though its value varies only between zero and 2 % of the total number of agents. The base value of this factor, representing the change in house ownership, is estimated by the researcher. It is also modelled in such a way that randomly selected household agents may change house ownership state every time step. Considering the influence of HO_{update} on the model output (given the current model conceptualization), it would be essential to acquire reliable data and better model representation of the factor to reduce the model output uncertainties.

The next influential factors are HI_{update}, HO_{ini}, and HI_{ini}. The base values of the household income-related factors are also based on our estimations as there is no publicly available record due to the sensitive nature of income data. Similarly, obtaining better dataset would help to reduce the output uncertainty. The initial house ownership variable is based on census data but agents' house ownership is assigned randomly as there is no available data regarding its spatial distribution. The $f_{abandoning}$ factor is influential in the case of primary measures abandoning as it sets a limit on the number of times an agent could abandon a measure. Better data would also reduce this factor's allocation to the model output uncertainty. All the other factors are non-influential as points representing these factors overlap around the $(0, 0)$ coordinate.

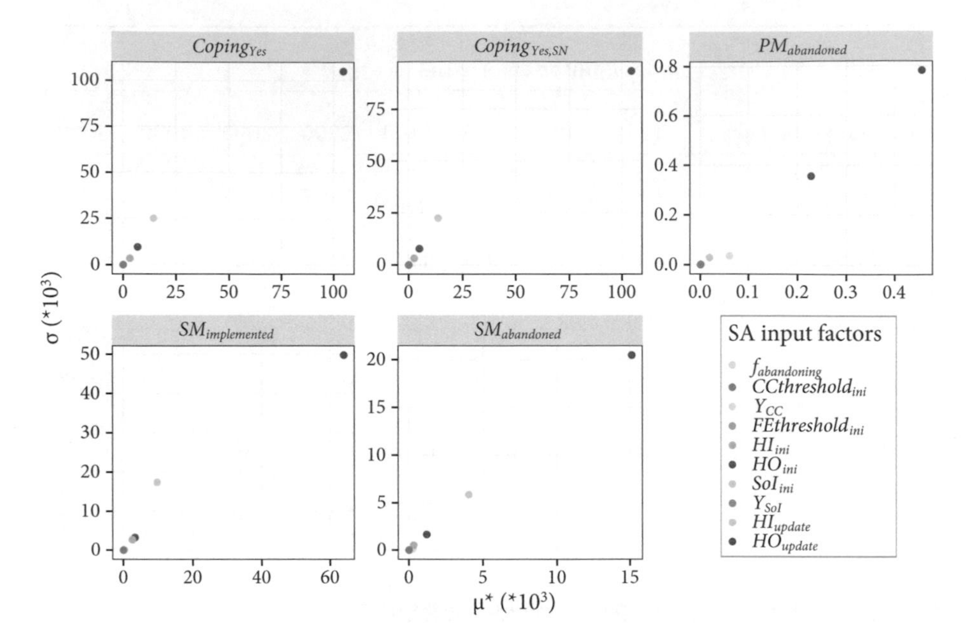

Figure 5.9 | Scatter plots displaying the Morris sensitivity measures μ^* and σ for five of the response factors. Points representing the least important factors may not be visible as they overlap close to the $(0, 0)$ coordinate.

5.5.3 Experimentation results

Effects of flood event scenarios

We have tested six different flood event scenarios, and the adaptation behaviours of agents are shown in Figure 5.10. The plots show that each scenario results in a unique trajectory of adaptation measures (see Figure 5.10(a)). However, Scenarios 1, 3, 4 and 6 have similar curves of $PM_{implemented}$ while Scenarios 1 and 4 appear to overlap. The two curves appear to overlap because the effect of the first event in Scenario 1 (Event B) is very small, and the second and biggest flood event (Event C) of Scenario 1, which happens at the same time as that of Scenario 4, dictates the number of measures implemented. Irrespective of the subsidy lever, the four scenarios have a similar number of $PM_{implemented}$ at the end of the simulation period. In these scenarios, the biggest event (Event C) occur as the first or the second event. As this event is big enough to flood every agent's house directly, most agents tend to develop protection motivation behaviour earlier. On the other hand, Scenarios 2 and 5 display a lower number of the response factor, which improves with a subsidy. In these scenarios, Event C occurs last; and hence, the $PM_{implemented}$ rises rapidly

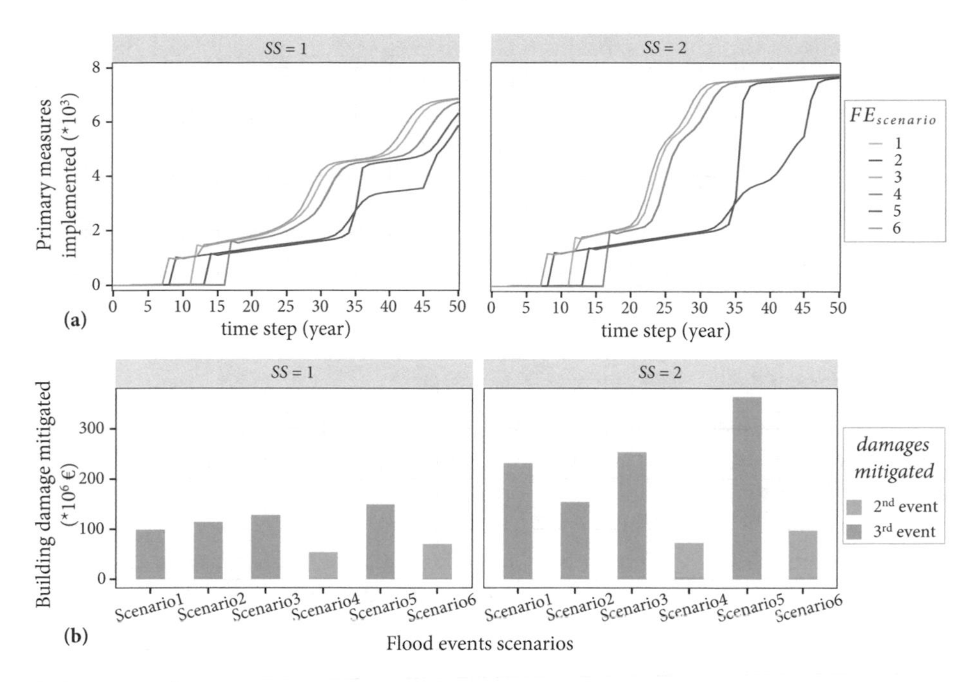

Figure 5.10 | Effects of six flood event scenarios on the adaptation behaviour of agents and the associated impact. (a) shows the cumulative number of primary measures implemented. In both plots, the curves for Scenarios 1 and 4 appear to overlap. (b) shows the potential building damage mitigated due to the primary measures implemented. In both (a) and (b), the left and right panels show the simulation results without subsidies and with subsidies for flooded agents, respectively.

after *time step* = 35. Furthermore, there are no major increases in the number of houses that implemented primary measures after the first flood events in the cases of Scenarios 1 and 2, i.e., after *time step* = 7 and *time step* = 2, respectively. The reason is that the first flood event in both scenarios (Event B) is a small event, and it only affects a few houses. Hence, its effect on the number of primary measures is minimal (but not zero). The curves appear flat, but there are minor increases in the slope of the curves after the mentioned time steps.

In terms of building damage mitigated, the scenarios with the two big events (C and A) occurring as first and second and within a short time interval display the least damage mitigated (see Figure 5.10(b) Scenarios 4 and 6). These are considered to be the worst cases of the six scenarios as agents did not have a coping behaviour before the first big event, and most agents did not yet develop coping behaviour when the second big event occurred after five years. Only 21 % and 14 % of the agents implemented a measure in cases of Scenarios 4 and 6, respectively, without subsidy. In contrast, in the case of Scenario 5, agents gradually develop coping behaviour after a first big event. By the time the second big event occurred after 37 years, about 45 % and 70 % of the agents already implemented a primary adaptation measure without subsidy and with a subsidy to flooded houses, respectively. Scenario 5 can be considered as the best scenario in which household agents have time to adapt and significantly reduce the potential damage that may occur in the future.

The main lesson from the results of the scenario exercise is that agents should be prepared or adapt quickly after an event to mitigate considerable potential damages. Big events may occur within a short time interval, and households should be prepared to mitigate associated damages. It should be noted that in Figure 5.10(b) there is no mitigated damage in the first event as we assumed that no mitigation measure was implemented initially.

Impacts of subsidies and shared strategies

The effects of the institutions are analysed in two categories. The first ones are the impacts of subsidies, and the second effects are that of the social network and shared strategy parameters.

Impacts of subsidies The cumulative number of implemented primary measures plotted in Figure 5.11 show that providing subsidies increases the protection motivation behaviour of agents irrespective of the flood event scenarios. For example, in the case of Scenario 1 flood event series, the building damage mitigated increases by about 130 % when a subsidy is provided to agents (see Figure 5.10(b)). However, giving subsidies either only to flooded agents or to all agents does not have a difference in the coping responses of agents. That is depicted by the overlapping curves of $SS = 2$ and $SS = 3$ in Figure 5.11. The result can be justified by the fact that (i) the subsidies only affect agents that implement permanent measures; and (ii) when a big flood event happens, it floods most of the agents, essentially levelling the number of agents impacted by $SS = 2$ and $SS = 3$.

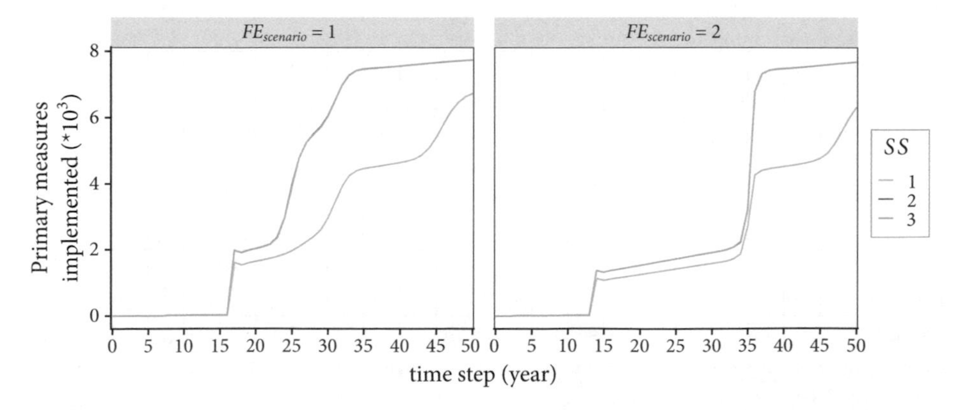

Figure 5.11 | Impacts of subsidy on the adaptation behaviour of agents. The subsidy levers 1, 2 and 3 represent no subsidy, subsidy only for flooded household agents and subsidy for all agents that consider flood as a threat, respectively. The left and right panels show simulation results with flood events scenarios of 1 and 2, respectively.

Impacts of social network and shared strategy parameters Figure 5.12 shows that an increase in the value of the social network parameter reduces the number of agents that develop a coping behaviour. As the SN parameter is associated with the proportion of coping agents within a house category, a higher SN requires a majority of agents in a given house category should have developed a coping behaviour to start influencing other agents. For example, when $SN = 0.5$, no agent is influenced by their social network as the criteria that at least 50 % of the agents in the same house category should have already implemented a measure to influence others has never been satisfied. On the other hand, when $SN = 0.2$, about 75 % of the agents that developed a coping behaviour after *time step* = 20 are influenced by their social network. Figure 5.12 also shows that the shared strategy parameter does not have a significant effect on the number of agents that develop a coping behaviour (for example, see the solid lines cluster together). This means that when the SN criteria are satisfied, most agents tend to follow the crowd.

In practical terms, this result shows that if agents need to wait to see many others implement measures to be influenced, most likely, they will not develop a motivation protection behaviour. Hence, aspects such as stronger community togetherness in which few neighbours can influence others to increase the possibility of implementing adaptation measures.

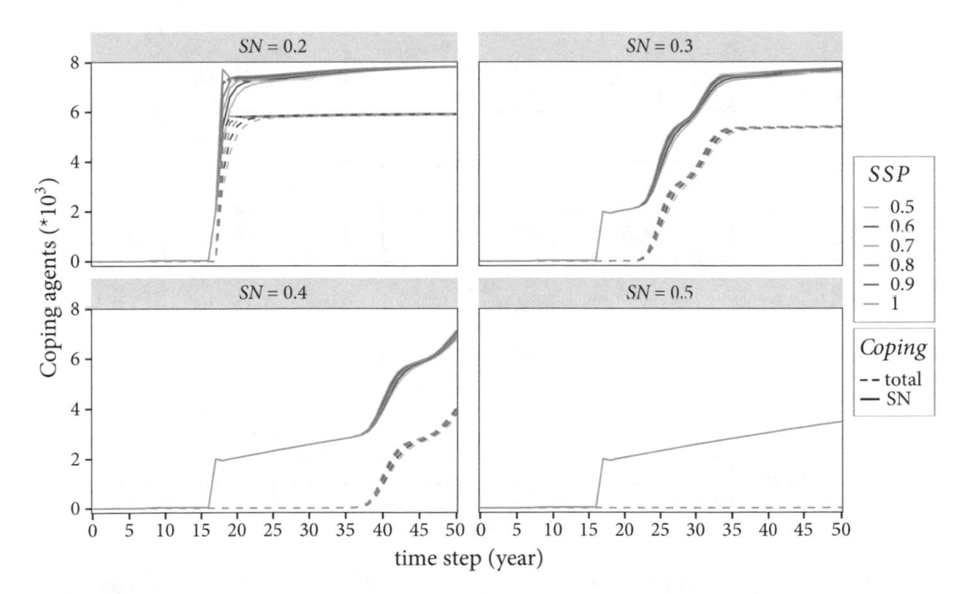

Figure 5.12 | Impacts of the social network and shared strategy parameter factors on the adaptation behaviour of agents. The solid lines show the total number of coping agents while the dashed lines show the agents that develop a coping behaviour influenced by their social network.

Impacts of individual strategies

In this section, we will analyse the effects of three factors that characterize individual strategies: delay parameter, adaptation duration and secondary measure parameter.

Impacts of delay parameter As shown in Figure 5.13, the percentage of agents that transform the coping behaviour to action decreases as the value of the delay parameter increases. When $DP = 1$, all agents that developed coping behaviour implement adaptation measures at the same time step. However, when $DP = 9$ (i.e., when the probability that a coping agent will implement a measure at a given year is 1/9), the number of agents that implement measures is 75 % of the number that develop a coping behaviour by the end of the simulation period.

Furthermore, both the number of coping agents and agents that implemented measures decreases with increase in DP value. For example, when $FE = 2$ and the value of DP increases from 1 to 9, the numbers of coping agents and agents that implemented a primary measure drop by about 27 % and 48 %, respectively, at *time step* = 50. This also has a knock-on effect on the implementation of a

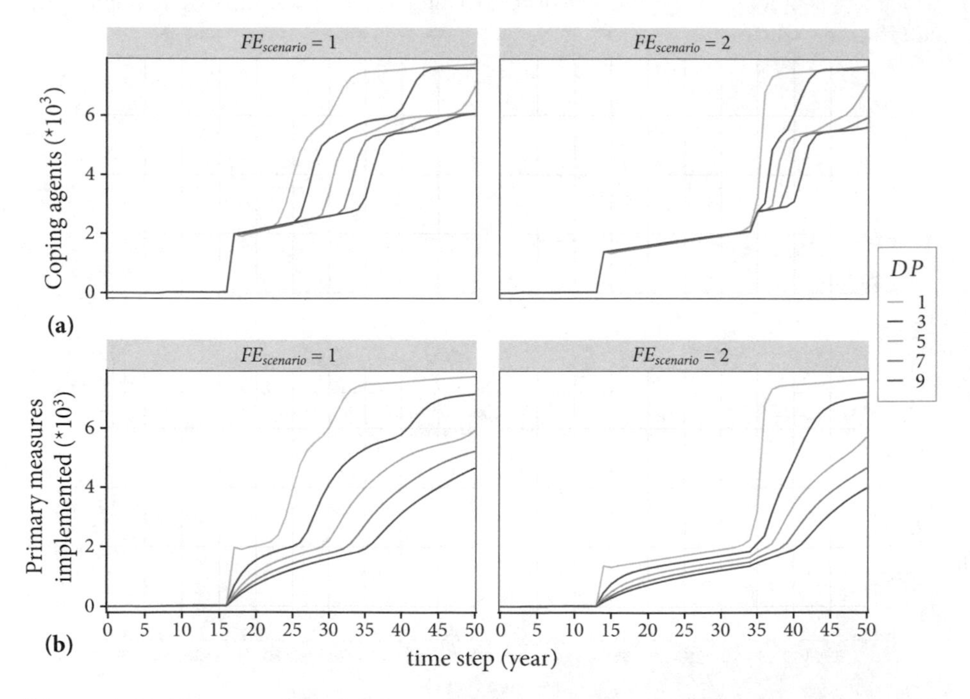

Figure 5.13 | Impacts of the delay parameter on the adaptation behaviour of agents. (a) shows the coping behaviour of agents and (b) shows the cumulative number of agents that converted their coping behaviour to action, i.e., implement primary adaptation measures. Simulations that generated the results are set with $SS = 2$. The left and right panels show simulation results with flood events scenarios of 1 and 2, respectively.

secondary measure, which reduces by about 50 %. Based on the outputs of the simulations, the delayed implementation of measures reduces the potential building and contents damage that could have been mitigated by €36.3 million and €8.7 million, respectively.

The main reason for the lower number of measures implemented with the increase in the value of the delay parameter is the decision of agents to delay the implementation. However, that also contributes to lower the number of agents influenced by their social network. In practical terms, this means that authorities should support households who tend to develop protection motivation behaviour so that they would implement adaptation measures promptly.

Impacts of adaptation duration parameter we evaluate the impacts of the adaptation duration using the number of agents that implemented and abandoned primary and secondary measures. The simulation results in Figure 5.14(a) show that the adaptation duration parameter has a minor impact on the number of primary and secondary measures implemented, regardless of the subsidy lever. For example, the largest percentage difference between the highest and lowest $PM_{implemented}$ is exhibited around *time step* = 30, which accounts about 28 %. One reason for the minor impact of $Y_{adaptation}$ on $PM_{implemented}$ could be that the parameter only affects agents that implement temporary primary measures, which is about half of the total number of agents. Another one could be that an increase in $PM_{implemented}$ also increases the number of agents that potentially abandon the measure. This is reflected in Figure 5.14(b) in which the peaks of $PM_{abandoned}$ correspond to the steepest slope of the curve displaying $PM_{implemented}$.

Figure 5.14(b) also shows that more agents abandon measures when the value of $Y_{adaptation}$ decreases. But then the number of measures abandoned decreases as agents reach the fixed number of times they could abandon measures, which is specified by the $f_{abandoning}$ parameter. In addition, the figure illustrates that, in general, $SM_{abandoned}$ is larger than $PM_{abandoned}$ along the simulation period. This can be explained by the model conceptualization, where agents first abandon secondary measures provided that they consider implementing them.

The practical lesson from the simulation results is that if agents tend to implement temporary measures, there should be a mechanism that encourages them to continue implementing the measures in future. For example, authorities may create and raise public awareness of how to seal windows and doors, and the availability of sandbags. This should be done regularly, and especially just before the event occurs as the measures can be implemented within a short period.

Impacts of secondary measure parameter Finally, we analyse the impacts of SMP on the number of agents that implemented secondary measures. Since the secondary measure conceptualized in the model is adapted furnishing, the effects of SMP are evaluated based on the contents damage mitigated.

Figure 5.15(a) shows that the cumulative number of agents that implemented secondary measure increases as the parameter value increases. But, the rate of in-

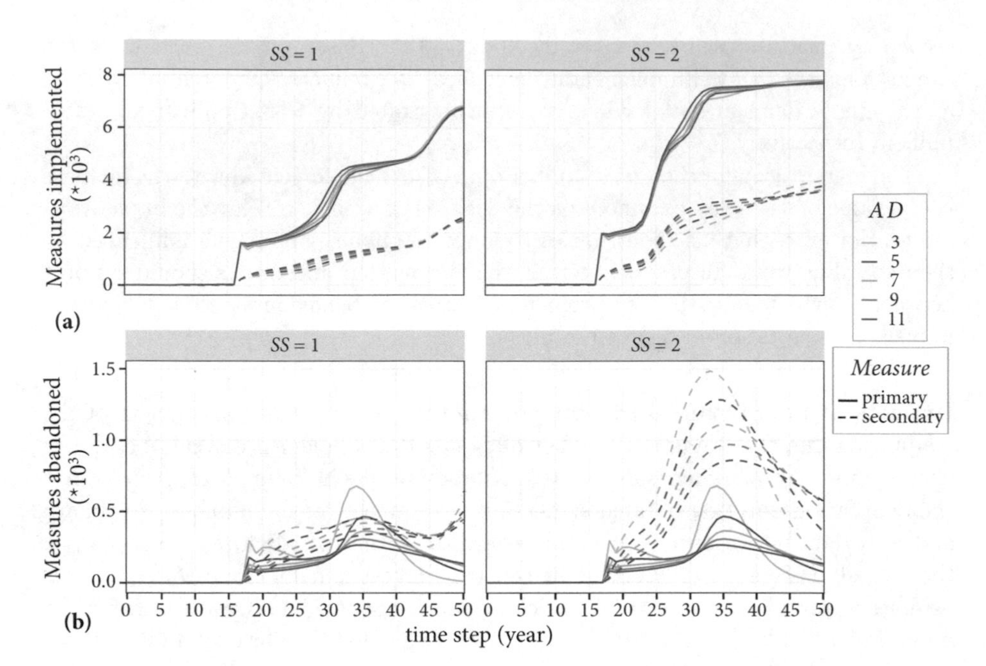

Figure 5.14 | Impacts of the adaptation duration on the adaptation behaviour of agents. (a) shows the primary and secondary measures implemented, and (b) shows the primary and secondary measures abandoned. The left and right panels show simulation results without subsidies and with subsidies for flooded agents, respectively.

crease in $SM_{implemented}$ is marginal especially after $SMP = 0.4$, in both cases of subsidy levers. When flooded agents receive a subsidy, $SM_{implemented}$ increases by about 1000 agents compared to the policy lever with no subsidy. Although the subsidy does not directly affect the implementation of secondary measures, it increases the implementation of primary measures, which in turn, increases $SM_{implemented}$. The only exception is when $SMP = 0$; in that case, no agent implement secondary measure despite the subsidy lever.

Similarly, Figure 5.15(b) shows that the contents damage mitigated increases marginally with the increase in the SMP value. The damage mitigated when $SMP = 0$ is because some agents implemented flood adapted interior fittings, which are classified as primary measures, and these measures mitigate both building and contents damages. When there is a subsidy, the contents damage mitigated increases by about three folds for each of the SMP values, except $SMP = 0$, compared to the policy lever with no subsidy.

The marginal increases in the $SM_{implemented}$ and the contents damage mitigated together with the increase in the values of SMP is because not all agents could implement secondary measures. As discussed in the model conceptualization, agents that live in bungalows and garden houses do not implement adapted furnishing since those house categories are single-storey houses. In general, based on our simulation

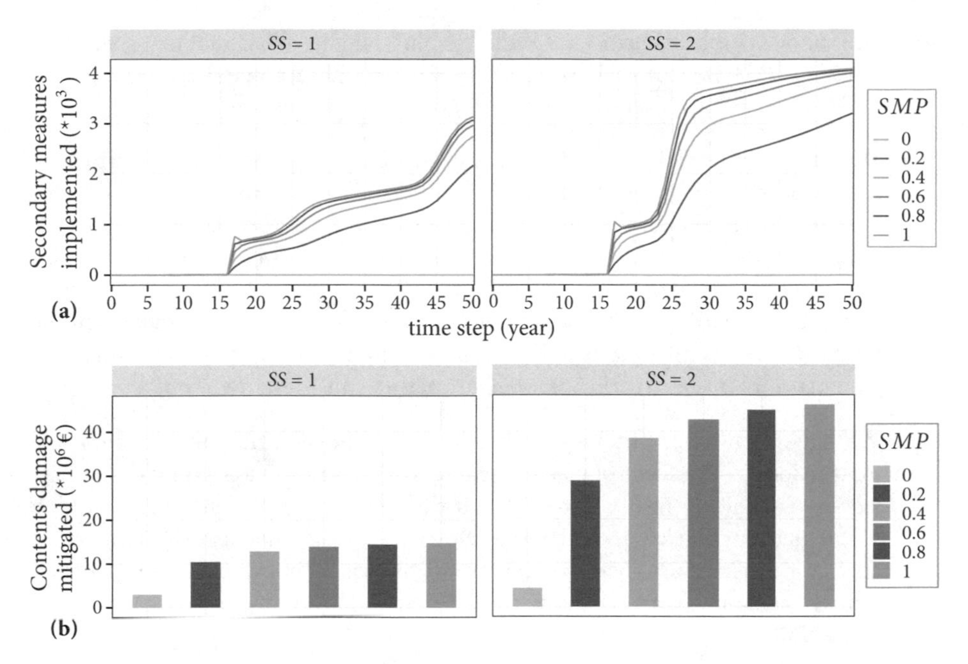

Figure 5.15 | Impacts of the secondary measure parameter on the adaptation behaviour of agents. (a) shows the cumulative number of secondary measures implemented, and (b) shows the potential contents damage mitigated. The left and right panels show simulation results without subsidies and with subsidies for flooded agents, respectively.

outputs, implementing only a secondary measure could mitigate more than €40 million. Hence, decision-makers should encourage households to consider implementing such simple measures that could be done at no monetary cost provided that there is space to keep contents safe.

5.6 Discussion and conclusion

The study aims to improve the current modelling practices of human-flood interaction and draw new insights for FRM policy design. Below, we discuss our modelling contributions and how they lead to policy insights.

i We have incorporated occurrences of flood events to examine how that influence household agents' adaptation behaviour. In our study, we examined six flood event scenarios, each comprising three coastal flood events occurring within 50 years simulation period. Simulation results show that a unique trajectory of adaptation measures and flood damages emerge from each flood event series. The interval between the occurrences of two big events is an important factor in defining households' adaptation behaviour. If a big event occurs first, it can serve as a wake-up call for future coping behaviours. However, that comes with a substantial amount of building and contents damage.

Households and authorities in Wilhelmsburg should avoid maladaptive practices (in PMT terms) such as avoidance and denial of possible future flooding and implement a measure to mitigate potential damages.

ii We have analysed the effects of a subsidy on the adaptation behaviour of individuals. We tested three subsidy levers: no subsidy, subsidy only for flooded household agents and subsidy for all agents that consider flood as a threat. Based on the simulation results, the last two levers have similar outcomes in terms of coping behaviours. It may depend on the flood event series, but providing subsidies increases the number of coping households in the long run. Hence authorities in Wilhelmsburg may consider providing subsidies to motivate households that implement permanent measures.

iii We have formulated the implementation of adaptation measures as informal institutions in the form of shared strategies that are influenced by social networks. Simulation results reveal that a wait-and-see approach, such as a high social network parameter settings, does not help to increase the number of coping households. There should be an approach in which fewer group of trusted community members or public figures may influence others in their community.

iv We have also analysed the effect of individual strategies on household adaptation behaviour. The strategies are delaying the implementation of measures, decisions on the adaptation durations of temporary measures and implementing secondary measures. Simulation results show that delaying measures implementation reduces millions of Euro that could have been mitigated. On the other hand, the overall impact of longer adaptation duration by some households could be cancelled out by the decision to abandon measures by others. It is essential to raise awareness continuously so that households do not forget or abandon to implement temporary measures. The role of simple measures such as adapted furnishing, which do not incur any monetary cost, should also be highlighted as these measures could contribute to reducing millions of Euro of contents damages.

In conclusion, the chapter presented a coupled agent-based and flood models developed to evaluate the adaptation behaviour and decision making of households to implement vulnerability reduction measures in the Wilhelmsburg quarter of Hamburg, Germany. We have employed the CLAIM framework to conceptualize the agent-flood interaction in the coupled model, and the protection motivation theory (PMT) to study household flood preparedness behaviour. The model conceptualization has benefitted from the qualitative exploration of PMT carried out in the same study area. Adding local knowledge of FRM issues and using other data sources, we extended the previous work by developing a simulation model that could support decision-making. Furthermore, the study has extended other prior works (Abebe *et al.*, 2019b; Erdlenbruch and Bonté, 2018; Haer *et al.*, 2016) to study human-flood interaction better and to gain new policy insights. With all the extensions, we have demonstrated that coupled ABM and flood models, together with a behavioural model, can potentially be used as decision support tools to examine the role of household adaptation measures in FRM. Although the focus of the chapter is the case of Wilhelmsburg, the improved modelling approach can be applied to any case to test policy levers and strategies considering heterogeneous individual behaviours.

It is worth mentioning that the results and analysis of the model outputs are subject to the limitations of the model conceptualizations. The threat and coping appraisals are modelled using rule-based decision trees. These trees are simplified ones that show linear and deterministic decision-making process by individuals. Although abstraction is an essential aspect of modelling, we acknowledge that actual decisions to a protection motivation behaviour can be more complex. Despite the stochastic elements in the model that could have provided unexpected results, the linear and deterministic nature of the decision trees may contribute to expected findings, especially the general trend.

Additionally, we defined the configuration of the decision trees (i.e., the importance of the factors that affect the threat and coping appraisal of individuals) based on previous empirical researches that are conducted in other study areas. However, some other factors could have been more important in a different study area. Hence, testing different decision-tree configurations would account the uncertainties in the model conceptualization. The trees could also have feedback loops in which the outcomes of agents' threat and coping appraisals could influence back the attributes that result in the appraisals. Future researches may use intelligent decision-making models such as Bayesian Networks as in (Abdulkareem *et al.*, 2018). In the flood model, considering dyke breach and other flood events and flood event series could be relevant modelling exercises.

The model conceptualization and the results would benefit from further refinement to provide more accurate insights into policy design. For example, more representative datasets are needed to reduce the input factors uncertainty as indicated by the sensitivity analysis. In our model conceptualization, households implement specific measures based on the category of a house they occupy, as defined in the shared strategies. Those are expert-based hypothetical strategies that could have been defined otherwise. We defined the institutions as shared strategies to give agents an option whether to develop a protection motivation behaviour or not. In the study area, there are no formal institutions that oblige households to implement any adaptation measure. We assumed, introducing institutions as shared strategies would be a reasonable starting point for the study area. Thus, the modelling exercises and their outcomes should be seen as an effort (i) to advance the use of coupled ABM-flood models in FRM, and (ii) to provoke communities and decision-makers in Wilhelmsburg to investigate further the role of household adaptation measures in mitigating potential damages. Furthermore, it is important to note that while the existing work addressed household measures, the same approach can also be applied to a range of different measures and contexts such as local and regional measures, nature-based solutions and traditional "grey infrastructure" which we intend to address in our future work.

Finally, the research presented can be enhanced by analysing model uncertainty. One may conceptualize the ABM differently, and investigating the impact of the different model conceptualization would be essential to communicate the uncertainty in model results. The research objective could also be extended by including other types of agents such as businesses and industries, and other response factors such as indirect damages (e.g., lost revenues due to business interruptions) to provide a broader view of the role of individual adaptation measures.

6

Insights into conceptualizing and modelling human-flood interactions

6.1 Introduction

This chapter addresses the insights gained from applying the CLAIM framework, the coupled ABM-flood models that were built using the framework, and the case studies. First, it gives concluding remarks about CLAIM. It also discusses the benefits and limitations of the framework and the associated modelling methodology, which details the steps to develop coupled ABM-flood models. Second, it highlights the insights gained from the use of CLAIM in conceptualizing and modelling of the case studies. It compares the two case study applications and discusses the advantages and limitations of the conceptualization and modelling. Finally, the chapter discusses the insights into socio-hydrologic modelling and FRM in general.

6.2 CLAIM and modelling methodology

Traditional FRM and flood modelling practices have been solely focusing on flood hazard reduction. However, as policymakers are challenged to develop resilient climate change adaptation and mitigation measures, impacts of these policies on the exposure and vulnerability of communities are increasingly important.

In this dissertation, we presented the CLAIM modelling framework, which allows for improved conceptualization and simulation of coupled human-flood systems. The human subsystem consists of heterogeneous agents and institutions that shape agents' decisions, actions and interactions, and are modelled using ABM. The flood subsystem consists of hydrologic and hydrodynamic processes that generate floods, and are modelled using numerical flood models. The dynamic link between the two subsystems happens through the urban environment.

The ABM is coupled with the flood model to study the behaviour (i.e., actions and interactions) of agents in relation to the defined institutions and to evaluate agents' exposure and vulnerability as well as the flood hazard. The methodology presented to build a coupled model is designed considering long-term FRM plans instead of operational level, during-flood strategies. The output of the coupled model is a level of flood risk in terms of assessed impact. The impact can be measured by simply counting the number of flooded houses or by computing the direct and indirect damages in monetary values. The assessed impact is used as a proxy to measure the effectiveness of the institutions in the study area.

Advantages

As argued by Parker *et al.* (2002, p. 213), although integrated models may combine "old areas of science", they enable to research problems in "new, more holistic ways". By incorporating the five main elements of a coupled human-flood system (i.e., agents, institutions, urban environment, physical processes that generate flood and external factors), CLAIM provides a *holistic* conceptualization and modelling of the human-flood interaction. CLAIM enables to develop conceptual human-flood models *systematically*, specifying each element and drawing the relationships between the elements. The explicit conceptualization of agents and institutions in CLAIM provides a better representation of the complex human subsystem compared to the use of differential equations employed in recent literature (for example, see Di Baldassarre *et al.*, 2013). It also helps to directly explore the effects of policies such as flood zoning policies on agents behaviour and system-level FRM by setting different policy levers.

Furthermore, CLAIM is designed to be as *generic* as possible so that it does not constrain the conceptualization of a specific case study. The level of representation of each element during conceptualization varies based on the problem that is modelled, the modeller's knowledge of the case study and the availability of data, among other factors. CLAIM also provides *flexibility* in terms of model development as a modeller can use any hydrodynamic model and ABM development platform to develop the coupled model.

It is also possible to *explicitly* model the human and flood subsystems using knowledge from the respective domains, and link the two subsystems dynamically to study their interactions. The framework provides an *interdisciplinary* approach by allowing knowledge integrations from hydrologists/hydraulic engineers and social scientists. The coupled ABM-flood modelling method allows to study how levels of flood hazard (i.e., flood magnitude and extent), exposure (i.e., number of assets-at-risk) and vulnerability (i.e., propensity to be affected) change simultaneously with changes in human behaviour (i.e., policies and their implementations).

As demonstrated in the Sint Maarten and Wilhemsburg FRM cases, models developed using CLAIM can assess the dynamic impacts of proposed policies, taking into account imperfectly rational and heterogeneous responses of individuals to the policies. Model outputs allow policymakers in FRM decision making to adopt an appropriate adaptation measure that will reduce future flood risk. For example, in Sint Maarten, implementing hazard reduction measures reduce the number of flooded houses significantly. In addition, the government may need to improve its inspection and enforcement of the Building Ordinance as it has a wider effect over the island in reducing the vulnerability of households.

Further, by incorporating implementations that change flood hazard, exposure and vulnerability, coupled ABM-flood models which utilize the CLAIM framework allow to assess flood risk as a function of time. This provides a more comprehensive view of the flood risk than if it were calculated based on a particular, historical and fixed urban environment condition.

Limitations

The non-specific and flexible design of CLAIM can be a constraint that CLAIM does not provide particular theories, scales or methods to model the human and flood subsystems. For example, a protection motivation theory was applied to model household agents' decision making in the Wilhelmsburg model (Chapter 5) while a simplified randomly generated agent decision was implemented in the case of Sint Maarten (Chapter 4). The level of complexity in the model representation can be highly subjective. This may not facilitate model development when the modeller (or the modelling team) is either less experienced or very diverse.

CLAIM is designed to analyse long-term, strategic level institutions that are considered during the flood disaster recovery and prevention stages. Though this is a design choice, conceptualizing and modelling operational level institutions — concerning the flood disaster preparedness and response stages — using CLAIM require amending the framework. For example, to study the implications of operational level institutions immediately before and after a flood event, the system conceptualization changes as the focus shifts to institutions such as flood early warning and dissemination plans, and evacuation policies. The agents and their attributes will change considering individual agents, and their age, gender or education status become more significant at that scale. The structure of the coupled model also changes because the ABM and the flood model run simultaneously to evaluate agents' behaviour when the flood propagates.

Another limitation of CLAIM is that conceptualizing and modelling two complex subsystems, i.e., the human and flood subsystems that in turn comprise further complex subsystems, require a large amount of data. The inclusion of additional or nested subsystems requires a balance between better representation of a system (or "needed complexity") and building a very complicated model (Sivapalan and Blöschl, 2015; Voinov and Shugart, 2013). In addition to the large data sets needed to build each model, running simulations may require substantial computational resources. Urban flood modelling using 2D hydrodynamic models present a high computational cost due to the smaller time step required to overcome simulation instability. In the case of ABMs, the computational demand is related to a large number of repetitions and experiments required to evaluate the model setup.

Concerning the coupled ABM-flood modelling methodology proposed to model human-flood interactions using CLAIM, the coupled model suffers from the limitations of both the ABM and flood modelling techniques. An important issue in modelling human-flood interaction is the parametrization of human behaviour in ABMs (see Crooks and Heppenstall, 2012, for a comprehensive discussion on the limitations of ABM). In a more technical note, model input-output exchange may pose a challenge to model integration. ABMs can be coded to generate outputs in any format, but hydrodynamic modelling software systems have specific input-output formats that may be interpreted only by the software. The uncertainty of the coupled model also increases as the uncertainties of the individual models may propagate.

6.3 Conceptualization and model development in case studies

In the previous two chapters, we have applied CLAIM to develop coupled ABM-flood models and tested how institutions shape heterogeneous agents' behaviour towards reducing flood hazard, exposure and vulnerability. This section compares the CLAIM conceptualization and the coupled models developed in the two chapters. It provides the advantages and limitations of the conceptualizations and the models developed. Key conceptualization and modelling aspects of the two cases are summarized in Table 6.1.

Conceptualization and modelling in the case of Sint Maarten

In Chapter 4, we mainly examined formal institutions, in the form of an ordinance and two policies, in the case of the Sint Maarten FRM. Of the formal institutions, one is in the draft stage while the rest are in effect. We classified the institutions as rules although they do not have a well-defined sanctioning, or the sanctions are not enforced. We tested the impact of one informal institution as well, in the form of a shared strategy. The agents identified are households and the government agent. The formal institutions have an impact on the exposure and vulnerability of household agents. They shape the behaviour of individual agents either to avoid living in flood-prone areas or to implement adaptation measures that reduce their vulnerability when they are exposed to flood hazard. The role of the government agent is to enforce the implementation of the rules. On the other hand, the informal institution has an impact on the flood hazard. Hazard reducing measures are public measures implemented by the government. Overall, the institutions drive all aspects of the flood risk elements in the study area.

In this case, we used only one set of flood-generating rainfall and storm surge event series. The flood hazard is affected by the urban development occurring in the study area. The effect of institutions on reducing flood risk was evaluated based on the flood impact on the island, expressed by the number of flooded houses. The policy levers examined were related to enforcing the formal institutions. The results of the coupled ABM-flood model allowed to analyse the levels of rule enforcement that provide system-level flood risk reduction, and identify the institution that is

Table 6.1 | Comparison of conceptualization and modelling characteristics between the FRM cases of Sint Maarten and Wilhelmsburg.

Factors	Sint Maarten	Wilhelmsburg
Institutions	Three rules and one shared strategy	Four shared strategies and one norm
	Three existing and one proposed	All hypothetical
Types of agents	Individuals (households) and government	Individuals (households) and authority
Agents spatial representation in the ABM	Households — explicit Government — no spatial representation	Households — explicit Authority — no spatial representation
Flood risk component the institutions affect	Hazard, exposure and vulnerability	Vulnerability
Considers urban development?	Yes	No
Type of flood	Pluvial and coastal	Coastal
Flood impact assessment metric	Number of flooded houses	Potential direct damages (building and contents)
Agent decision-making model	Random (based on thresholds)	Protection motivation theory
Flood-generating event series	One series of events	Multiple series of events
Number of simulation repetitions	Modeller's estimation based on computational demand	Based on experimental error variance analysis
Sensitivity analysis	One-factor-at-a-time	Elementary effects (Morris method)

more effective in reducing the impacts of flooding on the island. The results also allowed to assess if ratifying the proposed policy produces a "desirable" outcome.

Conceptualization and modelling in the case of Wilhelmsburg

In Chapter 5, we investigated the impacts of informal institutions in the FRM case of Wilhelmsburg. All the institutions are hypothetical ones as there is no practice of implementing flood mitigation measures at the household level. We classified four institutions as shared strategies in which agents could be influenced by the actions of their neighbours in implementing flood mitigation measures. The agents identified are households and the state authority. The latter may provide subsidy to household agents to implement temporary adaptation measures. The institution that define the subsidy is classified as a norm. The measures considered in the case reduce the vulnerability of household agents and building and contents damages. Urban development scenarios were not included in the model conceptualization.

The study improved some of the conceptualization and modelling limitations noted in Chapter 4. For example, it tested the impacts of multiple flood events series on the adaptation pattern of household agents. The study also applied a behavioural theory to investigate household-level decision making to adopt mitigation measures against flood threats instead of assigning agents decisions randomly. Regarding model evaluations, the study applied variance analysis of the experimental error, using the coefficient of variation metrics, to estimate the number of repetitions of the ABM that provide a representative output. Further, the elementary effects sensitivity analysis method applied in the study is an improved OFAT method than the one applied in Chapter 4.

General case study limitations

One of the limitations of the application of CLAIM on the case studies is that the only external factor explicitly included in both cases is the source of flood. The model conceptualizations in both cases do not include external economic and political factors. In fact, these factors were identified during the very first conceptualization stages. However, the modeller excluded them from the final conceptualizations as data was limiting to initialize parameters that describe these factors in the ABMs. Parameterizing the factors with hypothetical values would further increase the uncertainty of the models.

Another limitation regarding the CLAIM application is that the *change* arrow from Agents to Institutions (see Figure 3.1) has not been implemented in any of the case studies. The change refers to amending an institution by revising its influence and sanctioning (for example, when a norm is formalized to be a rule). The change also refers to creating a new institution or removing one. In both cases, we did not find a document that shows such institutional changes that are relevant to long-term FRM.

It should be highlighted that the limitation not to include some elements of the CLAIM in the case studies in this dissertation does not diminish the relevance of incorporating the elements in the framework. As described in the previous section, CLAIM is designed to be as generic as possible. Every element of the framework may not be applicable to every case study. As a result, acknowledging the limitations, we recommend addressing the issue in future studies as described in the next chapter.

Regarding model implementation, a maximum of one flood event per time step was assumed in both case studies. The main issue in this regard is the difference in time scale between flood events and policy implementations. Flood events usually occur in a time scale of hours and days, and could occur multiple times in a year. In contrast, policy implementations such as flood mitigation measures have a longer time scale (for example, years). Hence, we assumed only one flood event per year in the models developed to overcome software implementation challenges.

6.4 SOCIO-HYDROLOGIC MODELLING AND FRM STUDIES

One of the challenges in socio-hydrology is developing common frameworks that specify a comprehensive set of elements, and relationships among them, to conceptualize and analyse socio-hydrologic problems (Konar *et al.*, 2019). The CLAIM framework is a step forward in addressing the challenge, specifically, in structuring and modelling the human-flood interaction. CLAIM consists of five general elements to describe human-flood interactions and sets out the relationships between the elements. It provides a methodology to uniformly structure and model FRM issues in different case studies. The coupled ABM-flood models in both the Sint Maarten and the Wilhelmsburg cases followed a similar conceptualization procedure using the framework. Moreover, CLAIM can be used as a prototype to conceptualize the interaction of humans with other hydro-meteorological hazards. For example, in a human-drought interaction study, one way of adapting CLAIM is by quantifying hydrological aspects such as rainfall, evapotranspiration, soil moisture content and groundwater level in the physical processes and by changing the environment to agricultural fields including irrigation schemes. The institutions addressed in such a case would be, for example, water use and water allocation policies, crop diversification policies and traditional rangeland management norms.

Most socio-hydrologic models presented in the literature are stylized models of coupled non-linear differential equations (Barendrecht *et al.*, 2017; Konar *et al.*, 2019). However, the stylized modelling approach commonly used in socio-hydrology is criticized as it does not account for the heterogeneity of actors (Konar *et al.*, 2019). CLAIM addresses this challenge by employing ABMs as the primary approach to model heterogeneous agents. Agents may differ in their type or role, for example, household agents and a government agent in the case of Sint Maarten have distinct roles as the first follow (or not) policies while the latter enacts policies. Agents may have the same type but with different states and behaviours. For example, in the case of Wilhelmsburg, one household may have a direct flood experience while the other has not, or one household may decide to implement a dry proofing adaptation measure while another decides to implement wet proofing measure.

CLAIM further enhances the conceptualization of the human subsystem by incorporating institutions, which are the drivers that influence agents' behaviours (i.e., agents' actions and interactions). In ABMs, agents are commonly referred to have "rules" that govern their behaviours. In CLAIM, the institutions are the main "rules" although individual strategies may also shape agents' actions and interactions. Using the MAIA meta-model, CLAIM incorporates the relevant policies, ordinances and strategies that shape agents' behaviours. This allows addressing particular socio-hydrological issues through informing policies and supporting decision making. Agent interactions, feedback and decision makings were captured within the ABM.

Looking at FRM studies, in particular, Barendrecht *et al.* (2017) found in their review that most coupled human-flood models are *descriptive models*, which aim to understand the human-flood interactions based on observations of existing system state. The Sint Maarten coupled ABM-flood model falls under this category of models. Based on the observed and potential policy compliance and enforcement levels, the coupled model mainly investigates the implications of existing and draft FRM policies. On the other hand, the Wilhelmsburg coupled ABM-flood model is a *prescriptive model*, which aims to examine the effect of multiple possible future decisions that minimize flood risk. The model explores the role of household adaptation measures. Further improving the model with better datasets and conceptualization, it could provide an input to the FRM efforts of authorities in Wilhelmsburg.

7

REFLECTIONS AND OUTLOOK

The previous chapter offered insights into the conceptualization and modelling of human-flood interactions using the CLAIM framework and the associated modelling methodology. After summarizing the outcomes of the research, this chapter provides reflections on the research process that leads to this dissertation. These are personal reflections of the researcher regarding topics such as model development, model communication, learning new methodologies and interdisciplinary research. The chapter ends by giving an outlook on future research.

7.1 RESEARCH OUTPUTS

The dissertation started by posing three research questions. Below, we discuss how the research outputs address the questions.

The first question was: Which elements should be included to conceptualize the human-flood interaction? We have answered this question in Chapter 3. The CLAIM framework identifies five elements — agents, institutions, urban environment, physical processes and external factors. The first two are part of the human subsystem, while the physical processes define the flood subsystem. Depending on how they are specified, the external factors represent both subsystems. External economic and political factors are part of the human subsystem. In contrast, the sources of flooding are components of the flood subsystem. These components are the sole reason why flood-related disasters are called "natural disasters" as humans do not directly manage them. Indeed, if there is no flooding, there is no flood-related disaster. But, CLAIM clearly shows that the sources of flooding are just one side of the story. If people and their artefacts are not present where the flooding occurs, there would not also be a disaster. The urban environment element serves as a bridge between the two subsystems as both exist in this environment. It should be stressed that the five elements have broad interpretations that can be further specified based on the case studies (see for example Sections 4.3.1 and 5.4.1). Generally, we demonstrated that applying CLAIM facilitates decomposing and conceptualizing human-flood systems.

The second question was: How can we couple models that explicitly represent the human and the flood subsystems and the interactions between them? Throughout the dissertation, we have discussed and showed that coupled ABM-flood models explicitly represent human-flood interactions by applying interdisciplinary, domain knowledge. Emphasizing on human behaviour, ABMs simulate the actions, interactions and decision making of individuals and composite entities. Hydrodynamic 2D flood models are the conventional means of simulating urban flooding. In Section 3.3, we provided a generic methodology that allows building coupled ABM-flood models. The methodology follows standard model integration phases addressed in literature. In Chapters 4 and 5, we have applied the methods and demonstrated the details of the coupled model development in two case studies. The studies showed that the final model depends on the objectives of the modelling.

The third question was: How can coupled human-flood models that incorporate institutions such as risk drivers advance FRM? The commonly used terminology to describe the factors that affect agents behaviours in ABMs is "rules". However, in this study, we have adopted the well-founded and broader social science concept of institutions. In FRM context, the institutions are the formal and informal policies, conventions, agreements, norms and shared strategies that shape societies behaviours in mitigating or reducing flood risk. Hence, following or flouting what the institutions describe drives the flood hazard and communities' exposure and vulnerability. As shown in the case studies, human-flood interaction models that explicitly conceptualize institutions support policy decision making that, otherwise, would have not been achieved by the traditional flood modelling practices.

7.2 Reflections

On learning from the modelling process The value of modelling is not limited to analysing the model results. The modeller may also learn from the modelling process, including during data collection and conceptualization. For example, in the case of Wilhelmsburg, the authorities regularly send flyers to households regarding evacuation information in case flooding occurs. But, the same authorities claim that there is no need for households to implement individual adaptation measures because the public flood protections are built with high standards. Thus, they do not communicate risk and coping related message to the public as they feel that such a message may reduce residents' sense of security. However, one may argue that a flood event that triggers an evacuation will certainly have an impact on buildings and cause direct and indirect damages. Therefore, flood risk awareness campaigns should candidly present the probability of occurrence of a flood event and all associated risk (in terms of physical damage and risk to life) together with the uncertainty. They may also include the implementation of individual measures in their climate adaptation plans acknowledging that the public protection measures will not protect all scenarios of flood events.

On thinking outside the box In 2017, a wicked debate on "a tale of two disciplines: socio-hydrology and hydrosocial research" was organized at IHE Delft Institute for Water Education. A prominent promoter of socio-hydrologic research and speaker at the debate — Prof. Murugesu Sivapalan — mentioned that his hydrologist colleagues accused him of "socializing" the hydrological science. Another prominent scientist in the field of water resource systems engineering and management, Professor Daniel Loucks wrote: "In the recent past, some prominent hydrologists have resisted and objected to any inclusion of economic or social components linked to hydrologic processes" (Loucks, 2015, p. 4792). I have also experienced similar resistance from some colleagues with strong hydrology/hydraulics engineering background; or at least, they were not impressed by my effort to incorporate social science concepts in my research. As long as humans are impacted by or strive to manage water-related disasters, the two elements are interconnected. Hydrologists/hydraulics engineers should acknowledge this fact in their research, planning, management and engineering works. They need to work together with experts from the social sciences domain to address water-related challenges that need transdisciplinary perspective better. Including social-science courses in Hydrology and Hydraulic Engineering curriculums may also help future experts to recognize the need for transdisciplinary effort.

On ABM software implementation After developing conceptual models, converting such models to an ABM software is a daunting task. The main reason is the "nature" of the ABM modelling paradigm. In hydrodynamic modelling, 2D surface water flow in any case study can be modelled using the shallow water equations — a mass equation and two momentum equations in the x and y directions. If a modeller knows the initial conditions, the boundary conditions and the model parameter values of any study area, an off-the-shelf hydrodynamic modelling software such as MIKE21 solves the equations numerically and provide outputs such as water level and discharge at each computational cell. Unfortunately, there is no universal way of describing human behaviour in an ABM, especially considering heterogeneous agents and their interactions. In fact, there is no such ABM software. There are only ABM development environments such as NetLogo and Repast Simphony that provide the platform to write lines of codes that describe the conceptual model. Thus, developing ABM software requires a "certain" level of programming skills. In relation to that, using different ABM development platform requires knowledge of different programming languages. For example, NetLogo uses a simplified logo language while Repast Simphony uses the Java programming language. Besides, as every case is different, the modeller needs to develop the ABM software for every case. Researchers/modellers that will build ABMs should consider this fact while designing their research plan.

On the "subjectivity" of modelling Modelling human-water interactions certainly requires an understanding of the scientific knowledge such as the phenomena to be modelled (for example, a flood event and its impact), underlying physical laws and scientific concepts (for example, flow equations and protection motivation theory) and statistical methods to analyse model outputs. However, subjective considerations that are gained through observation and experience are also important. In coupled ABM-flood modelling, the level of subjective considerations while developing the individual models varies considerably. In numerical flood modelling, as there are governing shallow water equations that define 1D and 2D flows, the room for subjectivity is relatively low. Modellers may benefit from prior experience but model schematization, including time and space discretization, should satisfy stability conditions. In contrast, the ABM paradigm requires more personal judgement, creativity and imagination than hydrodynamic modelling. Starting from model conceptualization to software implementation and results analysis, ABMs are prone to the subjective interpretation of the modeller. An element one modeller considers as an important aspect can be conceptualized as an assumption by another, which would essentially create a different model. Hence, an ABM model development would benefit from a team of modellers and problem owners participating in every stage to reduce the subjective bias of a single modeller.

On model reproducibility and transparency When communicating their models, researchers tend to focus more on the results while not sufficiently giving significance to sharing the model itself. Openly communicating the model by clearly describing the modelling process (i.e., model conceptualization and all its assumptions), together with the result, is essential. Unfortunately, most researchers do not share models so that others could scrutinize or learn from them. I argue that sharing models should be a mainstream procedure for publication. In view of that, the software developed in this dissertation is accessible in GitHub for the sake of reproducibility and transparency. The programs may also be used as a starting point for other researchers who aspire to study human-flood interactions using coupled ABM-flood models. Complete lists of the assumptions made to develop the coupled models are available in Appendix A and Appendix B.

7.3 Outlook

The studies documented in this dissertation have imported social science concepts to the hydrological science. Institutions are the main ones in that regard. They are essential in human-flood systems modelling as they shape the actions and interactions within and between the subsystems. Institutions are one of the five elements that are explicitly defined in CLAIM. The modelling exercises presented in Chapter 4 and Chapter 5 analysed predefined institutions that agents may follow given their resources. Those ordinances and policies were conceptualized as fixed, and agents could not change or remove them over the computation duration. In reality, however, institutions are subject to change or could be eliminated if their impacts are deemed unsatisfactory. Institutions could also evolve as, for example, norms become rules. Future studies may extend the applicability of coupled ABM-flood models in socio-hydrologic studies by including endogenous institutional changes or the evolution of institutions through feedback mechanisms (as described in Ghorbani and Bravo, 2016; Smajgl *et al.*, 2008). Incorporating such institutional changes may provide insight into how FRM policies emerge bottom-up or evolve in the future.

Another concept we imported in this dissertation is the protection motivation theory (PMT). From a socio-hydrology perspective, the social part is the one that is "newly" introduced to the hydrological science. Therefore, there is a knowledge gap in modelling the social system. Applying the PMT in Chapter 5 enabled us to conceptualize and model agents' decision making better. But, more research is needed to test other behavioural models in socio-hydrology. To that end, psychologists, anthropologists and social geographers taking part in the modelling process would improve the model conceptualization.

In this dissertation, we developed models that focus only on long-term institutions, which influence FRM in the mitigation and recovery phases. Others developed ABMs to addressed operational-level human-flood interactions just before or immediately after a flood event (for example, Dawson *et al.*, 2011; Liu and Lim, 2018). I argue there should be a scientific curiosity, if not a practical one, to develop coupled ABM-flood models that examine institutions at every stage of the FRM phase. However, there are considerable challenges in developing these types of models. Conceptualizing and implementing such a model is a substantial undertaking, as there could be several types of agents and a lot of actions and interactions between agents. There would also be a time scale issue considering short term and long term institutional effects. Another major challenge is the availability of relevant data and computational resource. Nevertheless, if developed, such a model offers a more comprehensive insight into policy analysis and decision making.

The flood models presented in this study were developed using the MIKE Zero hydrodynamic software products. Unfortunately, 2D flood modelling using hydrodynamic software is a time-consuming process. One way of improving the issue of computational time could be using cellular automata (CA) inundation models which are faster than physically-based models (see, for example, Guidolin *et al.*, 2016). As CA models and ABMs have similarities (and even overlaps), coupling the two models would give a different dimension from the model implementation point of view. Besides the computational advantage, developing coupled ABM-CA models may smoothen the input-output data exchange between the component models.

As identified during model development, availability of data is a major limiting factor in human-flood interaction studies. In fact, hydrological and topographic data are more publicly available although sometimes in coarse resolution. In contrast, socio-economic data relevant to develop ABMs are scarce. Socio-hydrologic models, in general, will benefit from improved quality and quantity of datasets. Investing in such data and making it publicly available will repay as more and more knowledge will be generated. Modellers could also make use of the data to calibrate and validate their models. It allows to reduce the uncertainty of models and enhance the use of socio-hydrologic models for real-world policy analysis.

Appendix A

List of assumptions — Coupled ABM-flood model for Sint Maarten

To structure and conceptualize the Sint Maarten flood risk management case and develop the agent-based model, we have made the following assumptions. The reasons to make these assumptions are model simplification (i.e., to develop a less complicated model) and lack of data.

(a) All buildings have the same function, i.e., they are residential houses.

(b) Household agents are represented by the houses they live in; hence, they are static.

(c) There is a one-to-one relationship between household agents and houses (i.e., a household owns only one house and vice versa).

(d) Houses are geographically represented by a single point feature, which is the centroid of the house polygon. Houses are considered flooded if the point features intersect a flood extent map. This is a simplified way to compute impact. See (Chen *et al.*, 2016) that uses polygon features.

(e) All household agents know about all the institutions.

(f) If an agent decides to implement a measure or follow a policy, it implies that it has the financial resource to do so (for example, to elevate house or to upgrade the capacity of drainage channels).

(g) One type of structural measure is implemented in a catchment only once.

(h) Household agents do not implement hazard reduction measures. They only implement measures that reduce their vulnerability and exposure.

(i) If there is a decision to implement a structural measure, it will be implemented at the same time step. Its effect is evaluated in the next flood event.

(j) The government agent implements a structural measure only in one catchment per time step.

(k) Structural measures are designed for floods of rainfall with 50-year recurrence interval.

(l) Structural measures are implemented only after a flood event.

(m) The average lot size of a new house is 200 m^2. Hence, the average increase in CN value of a catchment for every new house built is 0.1. This does not consider other factors such as slope.

(n) The imperviousness of catchments is adjusted based only on the number of new houses built. We do not consider the expansion of roads, sidewalks and parking lots.

(o) A rainfall with a recurrence interval of 5-year is the minimum threshold that causes flooding. A rainfall magnitude below the 5-year recurrence interval does not result in flooding.

(p) Drainage channels in MIKE11 have the same roughness coefficient at every time step and in all the simulations (i.e., no blockage or special maintenance or cleaning is assumed).

(q) A maximum of one flood event occurs per time step.

(r) Rainfall is uniformly imposed on the study domain over the specified time period.

(s) No climate change impact considered. Design rainfall intensities and sea level are the same throughout the simulation period.

(t) MIKE21 is run with a hurricane-induced storm surge level of 0.5 m. This value does not change over time, and wave actions are not included.

(u) In the coupled model, flooding occurs after agent dynamics.

(v) A house is considered to be flooded if the flood depth is greater than 5 cm assuming that all houses have floor elevation of at least 5 cm.

(w) Only new houses apply measures such as elevated floors.

(x) Effect of policies and their implementations is evaluated based on the number of houses flooded. We neither considered other assets (e.g., flooded cars, boats and yachts) nor other impact metrics such as damages and business interruption losses in monetary values.

Appendix B

List of assumptions — Coupled ABM-flood model for Wilhelmsburg

To structure and conceptualize the Wilhelmsburg flood risk management case and develop the agent-based model, we have made the following assumptions. The reasons to make these assumptions are model simplification (i.e., to develop a less complicated model) and lack of data.

(a) Household agents are spatially represented by the houses they live in; hence, they are static.

(b) There is a one-to-one relationship between household agents and houses (i.e., a household owns only one house and vice versa).

(c) Houses are represented by polygon features such that each polygon represents one household agent. In the case of multi-storey buildings, the agent represents the household(s) living on the ground floor.

(d) When apartments and high-rise buildings are represented by one single polygon feature, the whole building is considered as one house representing one household agent.

(e) A maximum of one flood event occurs per time step.

(f) Only three flood event scenarios are considered. All the scenarios simulate dyke overtopping and have very low exceedance probability. Dyke breach is not considered in the conceptualization.

(g) When there is a flood, the flood depth of a house is extracted from the flood maps as the maximum of the flood depths read at the vertices of the polygon feature that represent the house.

(h) A house is considered to be flooded if the flood depth is greater than 10 cm assuming that all houses have floor elevation of at least 10 cm.

(i) Damage assessment does not include aspects such as damages on other assets (e.g., cars), indirect damage (e.g. business interruptions), risk to life, and structural collapse of buildings.

(j) Damage is assessed based only on the flood water level. The effect of floodwater velocity, duration and contamination level is not included in the damage assessment.

(k) Both building and content damages are assessed per building type. The damages of all houses of the same building type are calculated using the depth-damage curves for that building type.

(l) The sources of information does not initiate the coping appraisal process as in the original PMT study as agents know the kind of measure they implement.

(m) If a house has already appraised coping and implemented a measure, they do not appraise coping again, unless they abandon the measure, assuming that they do not implement another primary measure.

(n) Adaptation measures are sufficient to reduce flood damage in all flood events (perceived efficacy of measures).

(o) Agents are capable of successfully implementing adaptation measures (perceived self-efficacy).

(p) The effect of flood barriers such as flood protection walls and sandbags on the flood hydraulics is not accounted for.

(q) Agents only implement a maximum of one primary and one secondary measure at a given time step.

(r) Agents do not implement temporary adaptation measures (i.e., flood barriers) at any time step but deciding to implement the measures entails they only deploy them when there is a flood.

(s) If agents abandon measures, they only abandon non-permanent measures such as flood barriers.

(t) In case of non-permanent measures, if a household agent decides to implement a measure, the decision is valid at least for a year.

(u) If a household agent abandons a measure, it abandons it for at least a year.

(v) Household agents do not implement the same primary measure twice unless they abandon it.

(w) The adaptation duration specified in a simulation is the same for all temporary measures.

Appendix C

List of house types in Wilhelmsburg

EFH30A — Single-family house, Thermal insulation composite system

EFH30B — Single-family house, Cavity wall with insulation

EFH31A — Single-family house, plastered brick work, ground level: raised ground floor

EFH31B — Single-family house, plastered brick work, Souterrain/basement

EFH32A — Single-family house, plastered brick work

EFH32B — Single-family house, faced brick work

EFH34 — Single-family house, plastered brick work, Souterrain: apartment

EFH35A — Bungalow, plastered brick work

EFH35B — Bungalow, wooden construction

KGV33A — garden/summer house, plastered brick work

KGV33B — garden/summer house, wooden construction

MFH20A — Apartment building, basement: water proof concrete tanking

MFH21A — Apartment building, plastered brick work, ground level: apartments

MFH21B — Apartment building, faced brick work, ground level: apartments

MFH21C — Apartment building, faced reinforced concrete, ground level: apartments

MFH22A — Apartment building, faced brick work, ground level: business

MFH22B — Apartment building, faced brick work, ground level: business (same as MFH_22a)

MFH23A — Apartment building, plastered brick work, ground level: apartments

MFH23B — Apartment building, faced brick work, ground level: apartments

MFHH10 — High-rise building, dry construction, ground level: general use

MFHH11 — High-rise building, reinforced concrete, ground level: general use

MFHH12 — High-rise building, dry construction, ground level with garages

IGS — Hybrid house — IGS centre

OH — Hybrid house — Open house

HH — Hybrid house

SIG — Phase change material — smart is green

BIQ — Smart material house — BIQ

CS1 — Smart price house

GUS — Smart price house — Grundbau und Siedler (Do-it-yourself builders)

WH — Wälderhaus

WC — Wood Cube

Bibliography

Abar, S., G. K. Theodoropoulos, P. Lemarinier and G. M. P. O'Hare, 2017. "Agent Based Modelling and Simulation tools: A review of the state-of-art software." *Comput. Sci. Rev.* 24: 13–33. DOI: 10.1016/j.cosrev.2017.03.001.

Abdou, M., L. Hamill and N. Gilbert, 2012. "Designing and Building an Agent-Based Model." In A. J. Heppenstall, A. T. Crooks, L. M. See and M. Batty, eds., *Agent-Based Models of Geographical Systems*, pages 141–165. Springer, Dordrecht, The Netherlands.

Abdulkareem, S. A., E. W. Augustijn, Y. T. Mustafa and T. Filatova, 2018. "Intelligent judgements over health risks in a spatial agent-based model." *Int. J. Health Geographics* 17 (1): 8. DOI: 10.1186/s12942-018-0128-x.

Abebe, Y. A., A. Ghorbani, I. Nikolic, Z. Vojinovic and A. Sanchez, 2019a. "A coupled flood-agent-institution modelling (CLAIM) framework for urban flood risk management." *Environ. Model. Softw.* 111: 483–492. DOI: 10.1016/j.envsoft.2018.10.015.

Abebe, Y. A., A. Ghorbani, I. Nikolic, Z. Vojinovic and A. Sanchez, 2019b. "Flood risk management in Sint Maarten — A coupled agent-based and flood modelling method." *J. Environ. Manage.* 248: 109317. DOI: 10.1016/j.jenvman.2019.109317.

Aerts, J. C. J. H., W. J. Botzen, K. C. Clarke, S. L. Cutter, J. W. Hall, B. Merz, E. Michel-Kerjan, J. Mysiak, S. Surminski and H. Kunreuther, 2018. "Integrating human behaviour dynamics into flood disaster risk assessment." *Nat. Clim. Change* 8 (3): 193–199. DOI: 10.1038/s41558-018-0085-1.

Aguirre-Ayerbe, I., J. Martínez Sánchez, Í. Aniel-Quiroga, P. González-Riancho, M. Merino, S. Al-Yahyai, M. González and R. Medina, 2018. "From tsunami risk assessment to disaster risk reduction — the case of Oman." *Nat. Hazards Earth Syst. Sci.* 18 (8): 2241–2260. DOI: 10.5194/nhess-18-2241-2018.

Ahmed, E. and A. H. Hashish, 2006. "On modelling the immune system as a complex system." *Theory Biosci.* 124 (3-4): 413–418. DOI: 10.1016/j.thbio.2005.07.001.

Akhmadiyeva, Z. and I. Abdullaev, 2019. "Water management paradigm shifts in the Caspian Sea region: Review and outlook." *J. Hydrol.* 568: 997–1006. DOI: 10.1016/j.jhydrol.2018.11.009.

Allen, P. M., M. Strathern and J. Baldwin, 2008. "Complexity: the Integrating Framework for Models of Urban and Regional Systems." In S. Albeverio, D. Andrey, P. Giordano and A. Vancheri, eds., *The Dynamics of Complex Urban Systems: An Interdisciplinary Approach*, pages 21–41. Physica-Verlag, Heidelberg, Germany.

An, L., 2012. "Modeling human decisions in coupled human and natural systems: Review of agent-based models." *Ecol. Modell.* 229: 25–36. DOI: 10.1016/j.ecolmodel.2011.07.010.

APFM, 2012. *Urban Flood Management in a Changing Climate.* Flood Management Tools Series. Associated Programme on Flood Management (APFM), World Meteorological Organization, Geneva, Switzerland.

Axelrod, R., 2006. "Agent-based Modeling as a Bridge Between Disciplines." In L. Tesfatsion and K.L. Judd, eds., *Handbook of Computational Economics*, vol. 2, pages 1565–1584. Elsevier.

Ball, P., 2012. *Why society is a complex matter : Meeting twenty-first century challenges with a new kind of science.* Springer, Berlin, Heidelberg.

Bankes, S. C., 2002. "Agent-based modeling: A revolution?" *PNAS* 99 (suppl 3): 7199–7200. DOI: 10.1073/pnas.072081299.

Bar-Yam, Y., 1997. *Dynamics of complex systems.* Studies in Nonlinearity. Addison-Wesley, Reading, MA, USA.

Barendrecht, M. H., A. Viglione and G. Blöschl, 2017. "A dynamic framework for flood risk." *Water Secur.* 1: 3–11. DOI: 10.1016/j.wasec.2017.02.001.

Basurto, X., G. Kingsley, K. McQueen, M. Smith and C. M. Weible, 2010. "A Systematic Approach to Institutional Analysis: Applying Crawford and Ostrom's Grammar." *Polit. Res. Q.* 63 (3): 523–537. DOI: 10.1177/1065912909334430.

Behdani, B., 2012. "Evaluation of paradigms for modeling supply chains as complex socio-technical systems." In C. Laroque, J. Himmelspach, R. Pasupathy, O. Rose and A. Uhrmacher, eds., *Proceedings of the 2012 Winter Simulation Conference (WSC)*, pages 1–15. Berlin, Germany. DOI: 10.1109/WSC.2012.6465109.

Belete, G. F., A. Voinov and G. F. Laniak, 2017. "An overview of the model integration process: From pre-integration assessment to testing." *Environ. Model. Softw.* 87 (Supplement C): 49–63. DOI: 10.1016/j.envsoft.2016.10.013.

Bettencourt, L. M. A., 2015. "Cities as Complex Systems." In B. A. Furtado, P. A. M. Sakowski and M. H. Tóvolli, eds., *Modeling Complex Systems for Public Policies*, pages 217–236. Institute for Applied Economic Research (IPEA), Brasília, Brazil.

Birkholz, S. A., 2014. *The prospect of flooding and the motivation to prepare in contrasting urban communities: A qualitative exploration of Protection Motivation Theory.* Ph.D. thesis, Cranfield University, Cranfield, UK.

Blair, P. and W. Buytaert, 2016. "Socio-hydrological modelling: a review asking "why, what and how?"." *Hydrol. Earth Syst. Sci.* 20 (1): 443–478. DOI: 10.5194/hess-20-443-2016.

Boccara, N., 2004. *Modeling complex systems.* Graduate Texts in Contemporary Physics. Springer, New York, NY, USA.

Boelens, R., 2014. "Cultural politics and the hydrosocial cycle: Water, power and identity in the Andean highlands." *Geoforum* 57: 234–247. DOI: 10.1016/j.geoforum.2013.02.008.

Bollinger, L. A., C. B. Davis and I. Nikolic, 2013. "An Agent-Based Model of a Mobile Phone Production, Consumption and Recycling Network." In K. H. van Dam, I. Nikolic and Z. Lukszo, eds., *Agent-Based Modelling of Socio-Technical Systems*, Agent-Based Social Systems, pages 221–243. Springer Netherlands, Dordrecht, The Netherlands.

Bollinger, L. A., I. Nikolic, C. B. Davis and G. P. J. Dijkema, 2015. "Multimodel Ecologies: Cultivating Model Ecosystems in Industrial Ecology." *J. Ind. Ecol.* 19 (2): 252–263. DOI: 10.1111/jiec.12253.

Bolton, R. and T. J. Foxon, 2015. "A socio-technical perspective on low carbon investment challenges—Insights for UK energy policy." *Environ. Innov. Soc. Transit.* 14: 165–181. DOI: 10.1016/j.eist.2014.07.005.

Bonabeau, E., 2002. "Agent-based modeling: Methods and techniques for simulating human systems." *PNAS* 99 (Supplement 3): 7280–7287. DOI: 10.1073/pnas.082080899.

Botzen, W. J. W., H. Kunreuther, J. Czajkowski and H. d. Moel, 2019. "Adoption of Individual Flood Damage Mitigation Measures in New York City: An Extension of Protection Motivation Theory." *Risk Anal.* 39 (10): 2143–2159. DOI: 10.1111/risa.13318.

Bouleau, G., 2014. "The co-production of science and waterscapes: The case of the Seine and the Rhône Rivers, France." *Geoforum* 57: 248–257. DOI: 10.1016/j.geoforum.2013.01.009.

Brouwers, L. and M. Boman, 2010. "A Computational Agent Model of Flood Management Strategies." In H. A. do Prado, A. J. Barreto Luiz and H. C. Filho, eds., *Computational Methods for Agricultural Research: Advances and Applications*, pages 296–307. IGI Global, Hershey, USA.

Bruch, E. and J. Atwell, 2015. "Agent-based Models in Empirical Social Research." *Sociol. Methods Res.* 44 (2): 186–221. DOI: 10.1177/0049124113506405.

Brufau, P. and P. Garcia-Navarro, 2000. "Two-dimensional dam break flow simulation." *Int. J. Numer. Methods Fluids* 33 (1): 35–57. DOI: 10.1002/(SICI)1097-0363(20000515)33:1<35::AID-FLD999>3.0.CO;2-D.

Bubeck, P., W. J. W. Botzen and J. C. J. H. Aerts, 2012. "A Review of Risk Perceptions and Other Factors that Influence Flood Mitigation Behavior." *Risk Anal.* 32 (9): 1481–1495. DOI: 10.1111/j.1539-6924.2011.01783.x.

Bubeck, P., W. J. W. Botzen, H. Kreibich and J. C. J. H. Aerts, 2013. "Detailed insights into the influence of flood-coping appraisals on mitigation behaviour." *Global Environ. Change* 23 (5): 1327–1338. DOI: 10.1016/j.gloenvcha.2013.05.009.

Campolongo, F., J. Cariboni and A. Saltelli, 2007. "An effective screening design for sensitivity analysis of large models." *Environ. Model. Softw.* 22 (10): 1509–1518. DOI: 10.1016/j.envsoft.2006.10.004.

Carter, N., A. Viña, V. Hull, W. McConnell, W. Axinn, D. Ghimire and J. Liu, 2014. "Coupled human and natural systems approach to wildlife research and conservation." *Ecol. Soc.* 19 (3). DOI: 10.5751/ES-06881-190343.

Chen, A. S., M. J. Hammond, S. Djordjević, D. Butler, D. M. Khan and W. Veerbeek, 2016. "From hazard to impact: flood damage assessment tools for mega cities." *Nat. Hazards* 82 (2): 857–890. DOI: 10.1007/s11069-016-2223-2.

Ciullo, A., A. Viglione, A. Castellarin, M. Crisci and G. D. Baldassarre, 2017. "Socio-hydrological modelling of flood-risk dynamics: comparing the resilience of green and technological systems." *Hydrol. Sci. J.* 62 (6): 880–891. DOI: 10.1080/02626667.2016.1273527.

Conrad, S. A. and D. Yates, 2018. "Coupling stated preferences with a hydrological water resource model to inform water policies for residential areas in the Okanagan Basin, Canada." *J. Hydrol.* 564: 846–858. DOI: 10.1016/j.jhydrol.2018.07.031.

Crawford, S. E. S. and E. Ostrom, 1995. "A Grammar of Institutions." *Am. Polit. Sci. Rev.* 89 (03): 582–600. DOI: 10.2307/2082975.

CRED and UNISDR, 2015. *The human cost of weather related disasters 1995-2015.* Tech. rep., Centre for Research on the Epidemiology of Disasters (CRED) and United Nations Office for Disaster Risk Reduction (UNISDR).

Crooks, A. and A. Heppenstall, 2012. "Introduction to Agent-Based Modelling." In A. Heppenstall, A. Crooks, L. See and M. Batty, eds., *Agent-Based Models of Geographical Systems*, pages 85–105. Springer, Dordrecht, The Netherlands.

Crooks, A. T. and C. J. E. Castle, 2012. "The Integration of Agent-Based Modelling and Geographical Information for Geospatial Simulation." In A. J. Heppenstall, A. T. Crooks, L. M. See and M. Batty, eds., *Agent-Based Models of Geographical Systems*, pages 219–251. Springer, Dordrecht, The Netherlands.

Dawson, R. J., R. Peppe and M. Wang, 2011. "An agent-based model for risk-based flood incident management." *Nat. Hazards* 59 (1): 167–189. DOI: 10.1007/s11069-011-9745-4.

Department for Communities and Local Government, 2007. *Improving the flood performance of new buildings: flood resilient construction.* RIBA Publishing, London, UK.

Department for Communities and Local Government, 2016. *Land use change statistics in England: 2015–2016.* Tech. rep., Department for Communities and Local Government, London, UK.

Detsis, V., H. Briassoulis and C. Kosmas, 2017. "The Socio-Ecological Dynamics of Human Responses in a Land Degradation-Affected Region: The Messara Valley

(Crete, Greece).” *Land* 6 (3): 45. DOI: 10.3390/land6030045.

DHI, 2017a. *MIKE11: A Modelling System for Rivers and Channels — Reference Manual.* Tech. rep., MIKE Powered by DHI, Hørsholm, Denmark.

DHI, 2017b. *MIKE21 Flow Model FM: Hydrodynamic Module User Guide.* Tech. rep., MIKE Powered by DHI, Hørsholm, Denmark.

Di Baldassarre, G., A. Viglione, G. Carr, L. Kuil, J. L. Salinas and G. Blöschl, 2013. “Socio-hydrology: conceptualising human-flood interactions.” *Hydrol. Earth Syst. Sci.* 17 (8): 3295–3303. DOI: 10.5194/hess-17-3295-2013.

Di Baldassarre, G., A. Viglione, G. Carr, L. Kuil, K. Yan, L. Brandimarte and G. Blöschl, 2015. “Debates—Perspectives on socio-hydrology: Capturing feedbacks between physical and social processes.” *Water Resour. Res.* 51 (6): 4770–4781. DOI: 10.1002/2014WR016416.

Dijkema, G. P. J., Z. Lukszo and M. P. C. Weijnen, 2013. “Introduction.” In K. H. van Dam, I. Nikolic and Z. Lukszo, eds., *Agent-Based Modelling of Socio-Technical Systems*, Agent-Based Social Systems, pages 1–8. Springer Netherlands, Dordrecht, The Netherlands.

Dixon, L., N. Clancy, B. Bender, A. Kofner, D. Manheim and L. Zakaras, 2013. *Flood insurance in New York City following Hurricane Sandy.* Rand Corporation, Santa Monica, CA, USA.

DRRC, 2011. *Caribbean Implementation of the Hyogo Framework for Action: Mid-term Review.* Tech. rep., Caribbean Risk Management Initiative — UNDP Cuba, Disaster Risk Reduction Centre, University of the West Indies, Kingston, Jamaica.

Drummond, M. A., R. F. Auch, K. A. Karstensen, K. L. Sayler, J. L. Taylor and T. R. Loveland, 2012. “Land change variability and human-environment dynamics in the United States Great Plains.” *Land Use Policy* 29 (3): 710–723. DOI: 10.1016/j.landusepol.2011.11.007.

Dubbelboer, J., I. Nikolic, K. Jenkins and J. Hall, 2017. “An Agent-Based Model of Flood Risk and Insurance.” *JASSS* 20 (1): 6. DOI: 10.18564/jasss.3135.

Elshafei, Y., M. Sivapalan, M. Tonts and M. R. Hipsey, 2014. “A prototype framework for models of socio-hydrology: identification of key feedback loops and parameterisation approach.” *Hydrol. Earth Syst. Sci.* 18 (6): 2141–2166. DOI: 10.5194/hess-18-2141-2014.

Erdlenbruch, K. and B. Bonté, 2018. “Simulating the dynamics of individual adaptation to floods.” *Environ. Sci. Policy* 84: 134–148. DOI: 10.1016/j.envsci.2018.03.005.

Essenfelder, A. H., C. D. Pérez-Blanco and A. S. Mayer, 2018. “Rationalizing Systems Analysis for the Evaluation of Adaptation Strategies in Complex Human-Water Systems.” *Earth's Future* 6 (9): 1181–1206. DOI: 10.1029/2018EF000826.

European Commission, 2007. "Directive 2007/60/EC of the European Parliament and of the Council on the assessment and management of flood risks."

Filatova, T., P. H. Verburg, D. C. Parker and C. A. Stannard, 2013. "Spatial agent-based models for socio-ecological systems: Challenges and prospects." *Environ. Model. Softw.* 45: 1–7. DOI: 10.1016/j.envsoft.2013.03.017.

Floyd, D. L., S. Prentice-Dunn and R. W. Rogers, 2000. "A Meta-Analysis of Research on Protection Motivation Theory." *J. Appl. Soc. Psychol.* 30 (2): 407–429. DOI: 10.1111/j.1559-1816.2000.tb02323.x.

Fonoberova, M., V. A. Fonoberov and I. Mezić, 2013. "Global sensitivity/uncertainty analysis for agent-based models." *Reliab. Eng. Syst. Saf.* 118: 8–17. DOI: 10.1016/j.ress.2013.04.004.

Fraser, A., 2016. *Risk Root Cause Analysis Paper for PEARL (Preparing for Extreme And Rare events in coastaL regions project): St Maarten, Dutch Caribbean.* Tech. Rep. Paper #74, Department of Geography, King's College London.

Fraser, A., S. Paterson and M. Pelling, 2016. "Developing Frameworks to Understand Disaster Causation: From Forensic Disaster Investigation to Risk Root Cause Analysis." *J. Extreme Events* 03 (02): 1650008. DOI: 10.1142/S2345737616500081.

Gaillard, J. C., 2010. "Vulnerability, capacity and resilience: Perspectives for climate and development policy." *J. Int. Dev.* 22 (2): 218–232. DOI: 10.1002/jid.1675.

Gersonius, B., C. Zevenbergen and S. van Herk, 2008. "Managing flood risk in the urban environment: linking spatial planning, risk assessment, communication and policy." In C. Pahl-Wostl, P. Kabat and J. Möltgen, eds., *Adaptive and Integrated Water Management: Coping with Complexity and Uncertainty*, pages 263–275. Springer-Verlag, Berlin, Germany.

Ghorbani, A., 2013. *Structuring Socio-technical Complexity: Modelling Agent Systems using Institutional Analysis.* Ph.D. thesis, Delft University of Technology, Delft, The Netherlands.

Ghorbani, A., P. Bots, V. Dignum and G. Dijkema, 2013. "MAIA: a Framework for Developing Agent-Based Social Simulations." *JASSS* 16 (2): 9. DOI: 10.18564/jasss.2166.

Ghorbani, A. and G. Bravo, 2016. "Managing the commons: a simple model of the emergence of institutions through collective action." *International Journal of the Commons* 10 (1): 200–219. DOI: 10.18352/ijc.606.

Gilbert, N. and P. Terna, 2000. "How to build and use agent-based models in social science." *Mind Soc.* 1 (1): 57–72. DOI: 10.1007/BF02512229.

Glaser, M., P. Christie, K. Diele, L. Dsikowitzky, S. Ferse, I. Nordhaus, A. Schlüter, K. Schwerdtner Mañez and C. Wild, 2012. "Measuring and understanding sustainability-enhancing processes in tropical coastal and marine social-ecological systems." *Curr. Opin. Environ. Sustain.* 4 (3): 300–308. DOI: 10.1016/j.cosust.2012.05.004.

Gordon, D. M., 2002. "The organization of work in social insect colonies." *Complexity* 8 (1): 43–46. DOI: 10.1002/cplx.10048.

Grahn, T. and H. Jaldell, 2019. "Households (un)willingness to perform private flood risk reduction — Results from a Swedish survey." *Saf. Sci.* 116: 127–136. DOI: 10.1016/j.ssci.2019.03.011.

Grimm, V., U. Berger, F. Bastiansen, S. Eliassen, V. Ginot, J. Giske, J. Goss-Custard, T. Grand, S. K. Heinz, G. Huse, A. Huth, J. U. Jepsen, C. Jørgensen, W. M. Mooij, B. Müller, G. Pe'er, C. Piou, S. F. Railsback, A. M. Robbins, M. M. Robbins, E. Rossmanith, N. Rüger, E. Strand, S. Souissi, R. A. Stillman, R. Vabø, U. Visser and D. L. DeAngelis, 2006. "A standard protocol for describing individual-based and agent-based models." *Ecol. Modell.* 198 (1-2): 115–126. DOI: 10.1016/j.ecolmodel.2006.04.023.

Grimm, V., U. Berger, D. L. DeAngelis, J. G. Polhill, J. Giske and S. F. Railsback, 2010. "The ODD protocol: A review and first update." *Ecol. Modell.* 221 (23): 2760–2768. DOI: 10.1016/j.ecolmodel.2010.08.019.

Grothmann, T. and F. Reusswig, 2006. "People at Risk of Flooding: Why Some Residents Take Precautionary Action While Others Do Not." *Nat. Hazards* 38 (1): 101–120. DOI: 10.1007/s11069-005-8604-6.

Guha-Sapir, D., P. Hoyois, P. Wallemacq and R. Below, 2016. *Annual Disaster Statistical Review 2016: The Numbers and Trends.* Tech. rep., CRED, Brussels, Belgium.

Guidolin, M., A. S. Chen, B. Ghimire, E. C. Keedwell, S. Djordjević and D. A. Savić, 2016. "A weighted cellular automata 2D inundation model for rapid flood analysis." *Environ. Model. Softw.* 84: 378–394. DOI: 10.1016/j.envsoft.2016.07.008.

Haer, T., W. J. W. Botzen and J. C. J. H. Aerts, 2016. "The effectiveness of flood risk communication strategies and the influence of social networks—Insights from an agent-based model." *Environ. Sci. Policy* 60: 44–52. DOI: 10.1016/j.envsci.2016.03.006.

Harden, C. P., 2012. "Framing and Reframing Questions of Human-Environment Interactions." *Ann. Assoc. Am. Geogr.* 102 (4): 737–747. DOI: 10.1080/00045608.2012.678035.

Heckbert, S., T. Baynes and A. Reeson, 2010. "Agent-based modeling in ecological economics." *Ann. N.Y. Acad. Sci.* 1185: 39–53. DOI: 10.1111/j.1749-6632.2009.05286.x.

Helmke, G. and S. Levitsky, 2004. "Informal Institutions and Comparative Politics: A Research Agenda." *Perspect. Polit.* 2 (4): 725–740. DOI: 10.1017/S1537592704040472.

Hodgson, G. M., 1988. *Economics and Institutions: A Manifesto for a Modern Institutional Economics.* Polity Press, Cambridge, UK.

Holland, J. H., 2006. "Studying Complex Adaptive Systems." *J. Syst. Sci. Complex* 19 (1): 1–8. DOI: 10.1007/s11424-006-0001-z.

Holland, J. H., 2014. *Complexity: A Very Short Introduction.* Oxford University Press, Oxford, United Kingdom.

IPCC, 2012. *Managing the risks of extreme events and disasters to advance climate change adaption: Special report of Working Group I and II of the Intergovernmental Panel on Climate Change.* Cambridge University Press, Cambridge, UK and New York, NY, USA.

IPCC, 2014a. *Climate Change 2014: Impacts, Adaptation, and Vulnerability. Part A: Global and Sectoral Aspects. Contribution of Working Group II to the Fifth Assessment Report of the Intergovernmental Panel on Climate Change.* Cambridge University Press, Cambridge, UK and New York, NY, USA.

IPCC, 2014b. *Climate Change 2014: Impacts, Adaptation, and Vulnerability. Part B: Regional Aspects. Contribution of Working Group II to the Fifth Assessment Report of the Intergovernmental Panel on Climate Change.* Cambridge University Press, Cambridge, UK and New York, NY, USA.

IPCC, 2014c. "Summary for policymakers." In C. Field, V. Barros, D. Dokken, K. Mach, M. Mastrandrea, T. Bilir, M. Chatterjee, K. Ebi, Y. Estrada, R. Genova, B. Girma, E. Kissel, A. Levy, S. MacCracken, P. Mastrandrea and L. White, eds., *Climate Change 2014: Impacts, Adaptation, and Vulnerability. Part A: Global and Sectoral Aspects. Contribution of Working Group II to the Fifth Assessment Report of the Intergovernmental Panel on Climate Change*, pages 1–32. Cambridge University Press, Cambridge, UK and New York, NY, USA.

Jenkins, K., S. Surminski, J. Hall and F. Crick, 2017. "Assessing surface water flood risk and management strategies under future climate change: Insights from an Agent-Based Model." *Sci. Total Environ.* 595 (Supplement C): 159–168. DOI: 10.1016/j.scitotenv.2017.03.242.

Jennings, N. R. and M. J. Wooldridge, 1998. "Applications of Intelligent Agents." In N. R. Jennings and M. J. Wooldridge, eds., *Agent Technology: Foundations, Applications, and Markets*, pages 3–28. Springer Berlin Heidelberg, Berlin, Heidelberg.

Jepson, W. and H. L. Brown, 2014. "'If No Gasoline, No Water': Privatizing Drinking Water Quality in South Texas Colonias." *Environ Plan A: Econ. Space* 46 (5): 1032–1048. DOI: 10.1068/a46170.

Jha, A. K., R. Bloch and J. Lamond, 2012. *Cities and Flooding: A Guide to Integrated Urban Flood Risk Management for the 21st Century.* The World Bank, Washington DC, USA.

Kelly, R. A., A. J. Jakeman, O. Barreteau, M. E. Borsuk, S. ElSawah, S. H. Hamilton, H. J. Henriksen, S. Kuikka, H. R. Maier, A. E. Rizzoli, H. van Delden and A. A. Voinov, 2013. "Selecting among five common modelling approaches for integrated environmental assessment and management." *Environ. Model. Softw.* 47: 159–181. DOI: 10.1016/j.envsoft.2013.05.005.

Konar, M., M. Garcia, M. R. Sanderson, D. J. Yu and M. Sivapalan, 2019. "Expanding the Scope and Foundation of Sociohydrology as the Science of Coupled Human-Water Systems." *Water Resour. Res.* 55 (2): 874–887. DOI: 10.1029/2018WR024088.

Koutiva, I. and C. Makropoulos, 2016. "Modelling domestic water demand: An agent based approach." *Environ. Model. Softw.* 79: 35–54. DOI: 10.1016/j.envsoft.2016.01.005.

Kravari, K. and N. Bassiliades, 2015. "A Survey of Agent Platforms." *JASSS* 18 (1): 11. DOI: 10.18564/jasss.2661.

Kreibich, H. and A. H. Thieken, 2009. "Coping with floods in the city of Dresden, Germany." *Nat. Hazards* 51: 423–436. DOI: 10.1007/s11069-007-9200-8.

Kroes, P., M. Franssen, I. van de Poel and M. Ottens, 2006. "Treating socio-technical systems as engineering systems: some conceptual problems." *Syst. Res. Behav. Sci.* 23 (6): 803–814. DOI: 10.1002/sres.703.

Kuil, L., G. Carr, A. Viglione, A. Prskawetz and G. Blöschl, 2016. "Conceptualizing socio-hydrological drought processes: The case of the Maya collapse." *Water Resour. Res.* 52 (8): 6222–6242. DOI: 10.1002/2015WR018298.

Laniak, G. F., G. Olchin, J. Goodall, A. Voinov, M. Hill, P. Glynn, G. Whelan, G. Geller, N. Quinn, M. Blind, S. Peckham, S. Reaney, N. Gaber, R. Kennedy and A. Hughes, 2013. "Integrated environmental modeling: A vision and roadmap for the future." *Environ. Model. Softw.* 39: 3–23. DOI: 10.1016/j.envsoft.2012.09.006.

Lawrence, M. B., L. A. Avila, J. L. Beven, J. L. Franklin, J. L. Guiney and R. J. Pasch, 2001. "Atlantic Hurricane Season of 1999." *Mon. Weather Rev.* 129 (12): 3057–3084. DOI: 10.1175/1520-0493(2001)129<3057:AHSO>2.0.CO;2.

Lawrence, M. B., B. M. Mayfield, L. A. Avila, R. J. Pasch and E. N. Rappaport, 1998. "Atlantic Hurricane Season of 1995." *Mon. Weather Rev.* 126 (5): 1124–1151. DOI: 10.1175/1520-0493(1998)126<1124:AHSO>2.0.CO;2.

Levin, S. A., 1998. "Ecosystems and the Biosphere as Complex Adaptive Systems." *Ecosystems* 1 (5): 431–436. DOI: 10.1007/s100219900037.

Li, W. and Y. Li, 2012. "Managing Rangeland as a Complex System: How Government Interventions Decouple Social Systems from Ecological Systems." *Ecol. Soc.* 17 (1). DOI: 10.5751/ES-04531-170109.

Ligmann-Zielinska, A., D. B. Kramer, K. Spence Cheruvelil and P. A. Soranno, 2014. "Using Uncertainty and Sensitivity Analyses in Socioecological Agent-Based

Models to Improve Their Analytical Performance and Policy Relevance." *PLoS One* 9 (10). DOI: 10.1371/journal.pone.0109779.

Linton, J., 2014. "Modern water and its discontents: a history of hydrosocial renewal." *WIREs Water* 1 (1): 111–120. DOI: 10.1002/wat2.1009.

Linton, J. and J. Budds, 2014. "The hydrosocial cycle: Defining and mobilizing a relational-dialectical approach to water." *Geoforum* 57: 170–180. DOI: 10.1016/j.geoforum.2013.10.008.

Liu, J., T. Dietz, S. R. Carpenter, M. Alberti, C. Folke, E. Moran, A. N. Pell, P. Deadman, T. Kratz, J. Lubchenco, E. Ostrom, Z. Ouyang, W. Provencher, C. L. Redman, S. H. Schneider and W. W. Taylor, 2007. "Complexity of Coupled Human and Natural Systems." *Science* 317 (5844): 1513–1516. DOI: 10.1126/science.1144004.

Liu, X. and S. Lim, 2018. "An agent-based evacuation model for the 2011 Brisbane City-scale riverine flood." *Nat. Hazards* 94 (1): 53–70. DOI: 10.1007/s11069-018-3373-1.

Lorscheid, I., B.-O. Heine and M. Meyer, 2012. "Opening the 'black box' of simulations: increased transparency and effective communication through the systematic design of experiments." *Comput. Math. Organ. Theory* 18 (1): 22–62. DOI: 10.1007/s10588-011-9097-3.

Loucks, D. P., 2015. "Debates—Perspectives on socio-hydrology: Simulating hydrologic-human interactions." *Water Resour. Res.* 51 (6): 4789–4794. DOI: 10.1002/2015WR017002.

Löwe, R., C. Urich, N. Sto. Domingo, O. Mark, A. Deletic and K. Arnbjerg-Nielsen, 2017. "Assessment of urban pluvial flood risk and efficiency of adaptation options through simulations — A new generation of urban planning tools." *J. Hydrol.* 550: 355–367. DOI: 10.1016/j.jhydrol.2017.05.009.

Ludy, J. and G. M. Kondolf, 2012. "Flood risk perception in lands "protected" by 100-year levees." *Nat. Hazards* 61 (2): 829–842. DOI: 10.1007/s11069-011-0072-6.

Macal, C. M. and M. J. North, 2010. "Tutorial on agent-based modelling and simulation." *J. Simul.* 4 (3): 151–162. DOI: 10.1057/jos.2010.3.

Mark, O., S. Weesakul, C. Apirumanekul, S. B. Aroonnet and S. Djordjević, 2004. "Potential and limitations of 1D modelling of urban flooding." *J. Hydrol.* 299 (3): 284–299. DOI: 10.1016/j.jhydrol.2004.08.014.

Markard, J., M. Suter and K. Ingold, 2016. "Socio-technical transitions and policy change—Advocacy coalitions in Swiss energy policy." *Environ. Innov. Soc. Transit.* 18: 215–237. DOI: 10.1016/j.eist.2015.05.003.

MDC, 2015. *Hurricanes and Tropical Storms in the Dutch Caribbean.* Tech. rep., Meteorological Department Curaçao, Willemstad, Curaçao.

Merz, B., S. Vorogushyn, U. Lall, A. Viglione and G. Blöschl, 2015. "Charting unknown waters—On the role of surprise in flood risk assessment and management." *Water Resour. Res.* 51 (8): 6399–6416. DOI: 10.1002/2015WR017464.

Mitchell, M., 2009. *Complexity: A Guided Tour.* Oxford University Press, New York, NY, USA.

Morris, M. D., 1991. "Factorial Sampling Plans for Preliminary Computational Experiments." *Technometrics* 33 (2): 161–174. DOI: 10.2307/1269043.

Munich RE, 2012. *50th Anniversary of the North Sea Flood of Hamburg.* Press Dossier, Munich RE.

Mustafa, A., M. Bruwier, P. Archambeau, S. Erpicum, M. Pirotton, B. Dewals and J. Teller, 2018. "Effects of spatial planning on future flood risks in urban environments." *J. Environ. Manage.* 225: 193–204. DOI: 10.1016/j.jenvman.2018.07.090.

Mycoo, M. and M. G. Donovan, 2017. *A Blue Urban Agenda: Adapting to Climate Change in the Coastal Cities of Caribbean and Pacific Small Island Developing States.* Inter-American Development Bank, Washington DC, USA.

NatCen Social Research, 2017. *Climate concern and pessimism: Examining public attitudes across Europe.* Tech. rep., NatCen Social Research, London, UK.

Naulin, M., A. Kortenhaus and H. Oumeraci, 2012. "Reliability analysis and breach modelling of sea/estuary dikes and coastal dunes in an integrated risk analysis." *Coast. Eng. Proc.* 1 (33): management.61. DOI: 10.9753/icce.v33.management.61.

Nikolai, C. and G. Madey, 2009. "Tools of the Trade: A Survey of Various Agent Based Modeling Platforms." *JASSS* 12 (2): 2. DOI: http://jasss.soc.surrey.ac.uk/12/2/2.html.

Nikolic, I., K. H. van Dam and J. Kasmire, 2013. "Practice." In K. H. van Dam, I. Nikolic and Z. Lukszo, eds., *Agent-Based Modelling of Socio-Technical Systems*, Agent-Based Social Systems, pages 73–137. Springer, Dordrecht, Dordrecht, The Netherlands.

Nikolic, I. and J. Kasmire, 2013. "Theory." In K. H. van Dam, I. Nikolic and Z. Lukszo, eds., *Agent-Based Modelling of Socio-Technical Systems*, Agent-Based Social Systems, pages 11–71. Springer, Dordrecht, Dordrecht, The Netherlands.

North, D. C., 1990. *Institutions, Institutional Change and Economic Performance.* The Political Economy of Institutions and Decisions. Cambridge University Press, New York, NY, USA.

North, M. J., N. T. Collier, J. Ozik, E. R. Tatara, C. M. Macal, M. Bragen and P. Sydelko, 2013. "Complex adaptive systems modeling with Repast Simphony." *Complex Adapt. Syst. Model.* 1 (1): 3. DOI: 10.1186/2194-3206-1-3.

North, M. J. and C. M. Macal, 2007. *Managing Business Complexity: Discovering Strategic Solutions with Agent-Based Modeling and Simulation.* Oxford University

Press, New York, 1st ed.

O'Connell, P. E. and G. O'Donnell, 2014. "Towards modelling flood protection investment as a coupled human and natural system." *Hydrol. Earth Syst. Sci.* 18: 155–171. DOI: 10.5194/hess-18-155-2014.

O'Keefe, P., K. Westgate and B. Wisner, 1976. "Taking the naturalness out of natural disasters." *Nature* 260 (5552): 566–567. DOI: 10.1038/260566a0.

Ostrom, E., 1990. *Governing the Commons: The Evolution of Institutions for Collective Action.* The Political Economy of Institutions and Decisions. Cambridge University Press, Cambridge, UK.

Ostrom, E., 2009. "A General Framework for Analyzing Sustainability of Social-Ecological Systems." *Science* 325 (5939): 419–422. DOI: 10.1126/science.1172133.

Ostrom, E., R. Gardner and J. Walker, 1994. *Rules, Games, and Common-pool Resources.* University of Michigan Press, Ann Arbor, MI, USA.

Pahl-Wostl, C., 2015. "Shaping Human-Environment Interactions." In *Water Governance in the Face of Global Change: From Understanding to Transformation*, Water Governance—Concepts, Methods, and Practice, pages 125–158. Springer, Cham, Switzerland.

Panebianco, S. and C. Pahl-Wostl, 2006. "Modelling socio-technical transformations in wastewater treatment—A methodological proposal." *Technovation* 26 (9): 1090–1100. DOI: 10.1016/j.technovation.2005.09.017.

Parker, P., R. Letcher, A. Jakeman, M. B. Beck, G. Harris, R. M. Argent, M. Hare, C. Pahl-Wostl, A. Voinov, M. Janssen, P. Sullivan, M. Scoccimarro, A. Friend, M. Sonnenshein, D. Barker, L. Matejicek, D. Odulaja, P. Deadman, K. Lim, G. Larocque, P. Tarikhi, C. Fletcher, A. Put, T. Maxwell, A. Charles, H. Breeze, N. Nakatani, S. Mudgal, W. Naito, O. Osidele, I. Eriksson, U. Kautsky, E. Kautsky, B. Naeslund, L. Kumblad, R. Park, S. Maltagliati, P. Girardin, A. Rizzoli, D. Mauriello, R. Hoch, D. Pelletier, J. Reilly, R. Olafsdottir and S. Bin, 2002. "Progress in integrated assessment and modelling." *Environ. Model. Softw.* 17 (3): 209–217. DOI: 10.1016/S1364-8152(01)00059-7.

Pettersen, K. A., N. McDonald and O. A. Engen, 2010. "Rethinking the role of social theory in socio-technical analysis: a critical realist approach to aircraft maintenance." *Cogn. Technol. Work* 12 (3): 181–191. DOI: 10.1007/s10111-009-0133-8.

Pinter, N., F. Huthoff, J. Dierauer, J. W. F. Remo and A. Damptz, 2016. "Modeling residual flood risk behind levees, Upper Mississippi River, USA." *Environ. Sci. Policy* 58: 131–140. DOI: 10.1016/j.envsci.2016.01.003.

Plate, E. J., 2002. "Flood risk and flood management." *J. Hydrol.* 267 (1): 2–11. DOI: 10.1016/S0022-1694(02)00135-X.

Poussin, J. K., W. J. W. Botzen and J. C. J. H. Aerts, 2014. "Factors of influence on flood damage mitigation behaviour by households." *Environ. Sci. Policy* 40: 69–77. DOI: 10.1016/j.envsci.2014.01.013.

Poussin, J. K., W. J. Wouter Botzen and J. C. J. H. Aerts, 2015. "Effectiveness of flood damage mitigation measures: Empirical evidence from French flood disasters." *Global Environ. Change* 31: 74–84. DOI: 10.1016/j.gloenvcha.2014.12.007.

Price, R. K. and Z. Vojinovic, 2008. "Urban flood disaster management." *Urban Water J.* 5 (3): 259–276. DOI: 10.1080/15730620802099721.

Pyka, A. and T. Grebel, 2006. "Agent-Based Modelling – A Methodology for the Analysis of Qualitative Development Processes." In F. C. Billari, T. Fent, A. Prskawetz and J. Scheffran, eds., *Agent-Based Computational Modelling*, Contributions to Economics, pages 17–35. Physica-Verlag HD, Germany.

Qureshi, Z. H., 2007. "A Review of Accident Modelling Approaches for Complex Socio-technical Systems." In *Proceedings of the Twelfth Australian Workshop on Safety Critical Systems and Software and Safety-related Programmable Systems - Volume 86*, SCS '07, pages 47–59. Australian Computer Society, Inc., Darlinghurst, Australia.

Railsback, S. F. and V. Grimm, 2012. *Agent-Based and Individual-Based Modeling: A Practical Introduction.* Princeton University Press, Princeton, New Jersey, USA.

Railsback, S. F., S. L. Lytinen and S. K. Jackson, 2006. "Agent-based Simulation Platforms: Review and Development Recommendations." *SIMULATION* 82 (9): 609–623. DOI: 10.1177/0037549706073695.

Rand, W., 2015. "Complex Systems: Concepts, Literature, Possiblities and Limitations." In B. A. Furtado, P. A. M. Sakowski and M. H. Tóvolli, eds., *Modeling Complex Systems for Public Policies*, pages 37–54. Institute for Applied Economic Research (IPEA), Brasília, Brazil.

Rogers, R. W., 1983. "Cognitive and physiological process in fear appeals and attitudes changer: A revised theory of protection motivation." In J. T. Cacioppo and R. E. Petty, eds., *Social psychophysiology: A sourcebook*, pages 153–176. Guilford Press, New York, NY, USA.

Rotmans, J. and M. Van Asselt, 1996. "Integrated assessment: A growing child on its way to maturity." *Clim. Change* 34 (3): 327–336. DOI: 10.1007/BF00139296.

Saltelli, A., M. Ratto, T. Andres, F. Campolongo, J. Cariboni, D. Gatelli, M. Saisana and S. Tarantola, 2008. *Global Sensitivity Analysis: The Primer.* John Wiley & Sons, Chichester, England.

Schanze, J., 2006. "Flood Risk Management—A Basic Framework." In J. Schanze, E. Zeman and J. Marsalek, eds., *Flood Risk Management: Hazards, Vulnerability and Mitigation Measures*, vol. Earth and Environmental Sciences of *NATO Science Series*, pages 1–20. Springer, Dordrecht, The Netherlands.

Schlef, K. E., L. Kaboré, H. Karambiri, Y. C. E. Yang and C. M. Brown, 2018.

"Relating perceptions of flood risk and coping ability to mitigation behavior in West Africa: Case study of Burkina Faso." *Environ. Sci. Policy* 89: 254–265. DOI: 10.1016/j.envsci.2018.07.013.

Schlüter, M., R. R. J. Mcallister, R. Arlinghaus, N. Bunnefeld, K. Eisenack, F. Hölker, E. J. Milner-Gulland, B. Müller, E. Nicholson, M. Quaas and M. Stöven, 2012. "New Horizons for Managing the Environment: A Review of Coupled Social-Ecological Systems Modeling." *Nat. Resour. Model.* 25 (1): 219–272. DOI: 10.1111/j.1939-7445.2011.00108.x.

Scott, W. R., 1995. *Institutions and Organizations.* Foundations for Organizational Science. SAGE Publications, Inc, Thousand Oaks, California, USA, 2nd ed.

Sivapalan, M., 2015. "Debates—Perspectives on socio-hydrology: Changing water systems and the "tyranny of small problems"—Socio-hydrology." *Water Resour. Res.* 51 (6): 4795–4805. DOI: 10.1002/2015WR017080.

Sivapalan, M. and G. Blöschl, 2015. "Time scale interactions and the coevolution of humans and water." *Water Resour. Res.* 51 (9): 6988–7022. DOI: 10.1002/2015WR017896.

Sivapalan, M., H. H. G. Savenije and G. Blöschl, 2012. "Socio-hydrology: A new science of people and water." *Hydrol. Process.* 26 (8): 1270–1276. DOI: 10.1002/hyp.8426.

Smajgl, A., L. R. Izquierdo and M. Huigen, 2008. "Modeling endogenous rule changes in an institutional context: the adico sequence." *Adv. Complex Syst.* 11 (02): 199–215. DOI: 10.1142/S021952590800157X.

Smith, A. and A. Stirling, 2010. "The Politics of Social-ecological Resilience and Sustainable Socio-technical Transitions." *Ecol. Soc.* 15 (1): 11.

Sobiech, C., 2012. *Agent-Based Simulation of Vulnerability Dynamics: A Case Study of the German North Sea Coast.* Springer, Berlin Heidelberg.

Sorg, L., N. Medina, D. Feldmeyer, A. Sanchez, Z. Vojinovic, J. Birkmann and A. Marchese, 2018. "Capturing the multifaceted phenomena of socioeconomic vulnerability." *Nat. Hazards* 92 (1): 257–282. DOI: 10.1007/s11069-018-3207-1.

Stanilov, K., 2012. "Space in Agent-Based Models." In A. J. Heppenstall, A. T. Crooks, L. M. See and M. Batty, eds., *Agent-Based Models of Geographical Systems*, pages 253–269. Springer, Dordrecht, The Netherlands.

STAT, 2017. *Statistical Yearbook 2017.* Tech. rep., Department of Statistics Sint Maarten, Philipsburg, Sint Maarten.

ten Broeke, G., G. van Voorn and A. Ligtenberg, 2016. "Which Sensitivity Analysis Method Should I Use for My Agent-Based Model?" *JASSS* 19 (1): 5. DOI: 10.18564/jasss.2857.

Teschner, N., D. E. Orenstein, I. Shapira and T. Keasar, 2017. "Socio-ecological research and the transition toward sustainable agriculture." *Int. J. Agric. Sustain.* 15 (2): 99–101. DOI: 10.1080/14735903.2017.1294841.

Tessone, C. J., 2015. "The Complex Nature of Social Systems." In B. A. Furtado, P. A. M. Sakowski and M. H. Tóvolli, eds., *Modeling complex systems for public policies*, pages 141–168. Institute for Applied Economic Research (IPEA), Brasília, Brazil.

Tonn, G. L. and S. D. Guikema, 2017. "An Agent-Based Model of Evolving Community Flood Risk." *Risk Anal.* 38 (6): 1258–1278. DOI: 10.1111/risa.12939.

Troy, T. J., M. Konar, V. Srinivasan and S. Thompson, 2015a. "Moving sociohydrology forward: a synthesis across studies." *Hydrol. Earth Syst. Sci.* 19 (8): 3667–3679. DOI: 10.5194/hess-19-3667-2015.

Troy, T. J., M. Pavao-Zuckerman and T. P. Evans, 2015b. "Debates—Perspectives on socio-hydrology: Socio-hydrologic modeling: Tradeoffs, hypothesis testing, and validation." *Water Resour. Res.* 51 (6): 4806–4814. DOI: 10.1002/2015WR017046.

Ujeyl, G. and J. Rose, 2015. "Estimating Direct and Indirect Damages from Storm Surges: The Case of Hamburg—Wilhelmsburg." *Coast. Eng. J.* 57 (1): 1540006-1–1540006-26. DOI: 10.1142/S0578563415400069.

UN General Assembly, 1994. *Report of Global Conference on the Sustainable Development of Small Island Developing States (Barbados Programme of Action).* Tech. Rep. A/CONF.167/9, Bridgetown, Barbados.

UNISDR, 2015. *Making Development Sustainable: The Future of Disaster Risk Management.* Global Assessment Report on Disaster Risk Reduction. United Nations Office for Disaster Risk Reduction, Geneva.

van Emmerik, T. H. M., Z. Li, M. Sivapalan, S. Pande, J. Kandasamy, H. H. G. Savenije, A. Chanan and S. Vigneswaran, 2014. "Socio-hydrologic modeling to understand and mediate the competition for water between agriculture development and environmental health: Murrumbidgee River basin, Australia." *Hydrol. Earth Syst. Sci.* 18 (10): 4239–4259. DOI: https://doi.org/10.5194/hess-18-4239-2014.

Verhoog, R., A. Ghorbani and G. P. J. Dijkema, 2016. "Modelling socio-ecological systems with MAIA: A biogas infrastructure simulation." *Environ. Model. Softw.* 81: 72–85. DOI: 10.1016/j.envsoft.2016.03.011.

Viero, D. P., G. Roder, B. Matticchio, A. Defina and P. Tarolli, 2019. "Floods, landscape modifications and population dynamics in anthropogenic coastal lowlands: The Polesine (northern Italy) case study." *Sci. Total Environ.* 651: 1435–1450. DOI: 10.1016/j.scitotenv.2018.09.121.

Viglione, A., G. Di Baldassarre, L. Brandimarte, L. Kuil, G. Carr, J. L. Salinas, A. Scolobig and G. Blöschl, 2014. "Insights from socio-hydrology modelling on dealing with flood risk — Roles of collective memory, risk-taking attitude and trust." *J. Hydrol.* 518, Part A: 71–82. DOI: 10.1016/j.jhydrol.2014.01.018.

Voinov, A. and H. H. Shugart, 2013. "'Integronsters', integral and integrated modeling." *Environ. Model. Softw.* 39 (Supplement C): 149–158. DOI: 10.1016/j.envsoft.2012.05.014.

Vojinovic, Z., 2015. *Flood Risk: The Holistic Perspective—From Integrated to Interactive Planning for Flood Resilience.* IWA Publishing, London, UK.

Vojinovic, Z. and M. B. Abbott, 2012. *Flood Risk and Social Justice: From Quantitative to Qualitative Flood Risk Assessment and Mitigation.* Urban Hydroinformatics series. IWA Publishing, London, UK, 1st edition ed.

Vojinovic, Z., Y. A. Abebe, R. Ranasinghe, A. Vacher, P. Martens, D. J. Mandl, S. W. Frye, E. van Ettinger and R. de Zeeuw, 2013. "A machine learning approach for estimation of shallow water depths from optical satellite images and sonar measurements." *J. Hydroinf.* 15 (4): 1408–1424. DOI: 10.2166/hydro.2013.234.

Vojinovic, Z., M. Hammond, D. Golub, S. Hirunsalee, S. Weesakul, V. Meesuk, N. Medina, A. Sanchez, S. Kumara and M. Abbott, 2016. "Holistic approach to flood risk assessment in areas with cultural heritage: a practical application in Ayutthaya, Thailand." *Nat. Hazards* 81 (1): 589–616. DOI: 10.1007/s11069-015-2098-7.

Vojinovic, Z. and J. Huang, 2014. *Unflooding Asia: The Green Cities Way.* IWA Publishing, London, UK.

Vojinovic, Z., S. D. Seyoum, J. M. Mwalwaka and R. K. Price, 2011. "Effects of model schematisation, geometry and parameter values on urban flood modelling." *Water Sci. Technol.* 63 (3): 462–467. DOI: 10.2166/wst.2011.244.

Vojinovic, Z. and J. van Teeffelen, 2007. "An integrated stormwater management approach for small islands in tropical climates." *Urban Water J.* 4 (3): 211–231. DOI: 10.1080/15730620701464190.

Votsis, A., 2017. "Utilizing a cellular automaton model to explore the influence of coastal flood adaptation strategies on Helsinki's urbanization patterns." *Comput. Environ. Urban Syst.* 64: 344–355. DOI: 10.1016/j.compenvurbsys.2017.04.005.

Watson, M., 2012. "How theories of practice can inform transition to a decarbonised transport system." *J. Transp. Geogr.* 24: 488–496. DOI: 10.1016/j.jtrangeo.2012.04.002.

Werner, B. T. and D. E. McNamara, 2007. "Dynamics of coupled human-landscape systems." *Geomorphology* 91 (3): 393–407. DOI: 10.1016/j.geomorph.2007.04.020.

Wescoat, J. L., A. Siddiqi and A. Muhammad, 2018. "Socio-Hydrology of Channel Flows in Complex River Basins: Rivers, Canals, and Distributaries in Punjab, Pakistan." *Water Resour. Res.* 54 (1): 464–479. DOI: 10.1002/2017WR021486.

Wesselink, A., M. Kooy and J. Warner, 2016. "Socio-hydrology and hydrosocial analysis: toward dialogues across disciplines." *WIREs Water* 4 (2): e1196. DOI: 10.1002/wat2.1196.

White, G. F., 1945. *Human Adjustment to Floods: A Geographical Approach to the Flood Problem in the United States.* Research Paper 29. University of Chicago, Chicago, IL, USA.

Wilensky, U. and W. Rand, 2015. *An Introduction to Agent-Based Modeling: Modeling Natural, Social, and Engineered Complex Systems with NetLogo.* The MIT Press, Cambridge, Massachusetts.

Wolfram, S., 2002. *A new kind of science.* Wolfram Media, Champaign, IL.

Xu, L., P. Gober, H. S. Wheater and Y. Kajikawa, 2018. "Reframing sociohydrological research to include a social science perspective." *J. Hydrol.* 563: 76–83. DOI: 10.1016/j.jhydrol.2018.05.061.

Young, O. R., 1986. "International Regimes: Toward a New Theory of Institutions." *World Politics* 39 (1): 104–122. DOI: 10.2307/2010300.

Yu, D. J., N. Sangwan, K. Sung, X. Chen and V. Merwade, 2017. "Incorporating institutions and collective action into a sociohydrological model of flood resilience." *Water Resour. Res.* 53: 1336–1353. DOI: 10.1002/2016WR019746.

Ziegler, J. P., E. J. Golebie, S. E. Jones, B. C. Weidel and C. T. Solomon, 2017. "Social-ecological outcomes in recreational fisheries: the interaction of lakeshore development and stocking." *Ecol. Appl.* 27 (1): 56–65. DOI: 10.1002/eap.1433.

About the Author

Yared Abayneh Abebe has received his Bachelor of Science degree (with distinction) in Soil and Water Engineering and Management from Haramaya University, Ethiopia in July 2006. After working in Jimma University, Ethiopia as Assistant Lecturer and Researcher for three years, he joined the Master of Science degree in Water Science and Engineering, specialization Hydroinformatics at IHE Delft Institute for Water Education, Delft, The Netherlands. Yared graduated (with distinction) in April 2011. His graduate research work focused on "Comparison of numerical schemes for modelling supercritical and transcritical flows along urban floodplains". The graduate study was funded by the NUFFIC-NFP Fellowship programme.

After earning his diploma, he worked on a project that focused on flood risk reduction in Small Island Developing States. His main achievements include developing nearshore bathymetry using satellite and sonar data, modelling coastal and inland flooding using numerical modelling software; assessing flood hazard, vulnerability and risk; and evaluating different structural and non-structural flood risk reduction measures.

Yared started his PhD research in November 2013 in the Urban Water Systems group of the Environmental Engineering and Water Technology Department at IHE Delft. The PhD study is funded by the European Union's FP7 project PEARL (Preparing for Extreme And Rare events in coastaL regions) and Horizon 2020 project RECONECT (Regenerating ECOsystems with Nature-based solutions for hydro-meteorological risk rEduCTion).

In 2017, Yared was part of a PEARL project team that went for a fact-finding and needs assessment mission to Sint Maarten after the island was struck by the deadly Hurricane Irma on 6 September 2017. In the mission, the team conducted workshops, interviews and household surveys regarding hurricane warnings, evacuations, hurricane preparedness, people's awareness/perception of hurricane impacts and organizational responses after a hurricane.

List of Publications

Peer-reviewed publications

Abebe, Y.A., Ghorbani, A., Nikolic, I., Manojlovic, N., Gruhn, A. and Vojinovic, Z. The role of household adaptation measures to reduce vulnerability to flooding: a coupled agent-based and flood modelling approach. *Hydrol. Earth Syst. Sci. Discuss. (Accepted for publication).* DOI: https://doi.org/10.5194/hess-2020-272

Medina, N., Abebe, Y.A., Sanchez, A. and Vojinovic, Z. (2020). Surveying after a disaster. Assessing Socioeconomic Vulnerability after a Hurricane: A Combined Use of an Index-Based approach and Principal Components Analysis. *Sustainability.* 12(04): 1452. DOI: https://doi.org/10.3390/su12041452

Medina, N., Abebe, Y.A., Sanchez, A., Vojinovic, Z. and Nikolic, I. (2019). Surveying after a disaster. Capturing Elements of Vulnerability, Risk and Lessons Learned from a Household Survey in the Case Study of Hurricane Irma in Sint Maarten. *Journal of Extreme Events.* 06(02): 1950001. DOI: https://doi.org/10.1142/S2345737619500015

Abebe, Y.A., Ghorbani, A., Vojinovic, Z., Nikolic, I. and Sanchez, A. (2019). Flood risk management in Sint Maarten — A coupled agent-based and flood modelling method. *Journal of Environmental Management.* 248, 109317. DOI: https://doi.org/10.1016/j.jenvman.2019.109317

Pyatkova, K., Chen, A.S., Djordjevic, S., Butler, D., Vojinovic, Z., Abebe, Y.A. and Hammond, M. (2019). Flood Impacts on Road Transportation Using Microscopic Traffic Modelling Techniques. In: Behrisch M., Weber M. (eds) Simulating Urban Traffic Scenarios. *Lecture Notes in Mobility.* Springer, Cham, Switzerland. pp 115-126. DOI: https://doi.org/10.1007/978-3-319-33616-9

Abebe, Y.A., Ghorbani, A., Nikolic, I., Vojinovic, Z. and Sanchez, A. (2019). A coupled flood-agent-institution modelling (CLAIM) framework for urban flood risk management. *Environmental Modelling & Software.* 111, 483-492. DOI: https://doi.org/10.1016/j.envsoft.2018.10.015

Abebe, Y.A., Seyoum, S.D., Vojinovic, Z. and Price, R.K. (2016). Effects of reducing convective acceleration terms in modelling supercritical and transcritical flow conditions. *Water.* 8(12):562. DOI: https://doi.org/10.3390/w8120562

Vojinovic, Z., Abebe, Y.A., Ranasinghe, R., Vacher, A., Martens, P., Mandl, D.J., Frye, S.W., van Ettinger, E. and de Zeeuw, R. (2013). A machine learning approach for estimation of shallow water depths from optical satellite images and sonar measurements. *Journal of Hydroinformatics.* 15(4):1408-1424. DOI: https://doi.org/10.2166/hydro.2013.234

Vojinovic, Z., Seyoum, S., Salum, M.H., Price, R.K., Fikri, A.F. and Abebe, Y. (2013). Modelling floods in urban areas and representation of buildings with a method based on adjusted conveyance and storage characteristics. *Journal of Hydroinformatics.* 15(4):1150-1168. DOI: https://doi.org/10.2166/hydro.2012.181

Conference proceedings

Abebe, Y.A., Ghorbani, A., Vojinovic, Z., Nikolic, I. and Sanchez, A. (2016). Agent-based simulation for flood risk assessment. *12th International Conference on Hydroinformatics (HIC 2016)*, 21-26 August, Incheon, Korea.

Abebe, Y.A., Ghorbani, A., Vojinovic, Z., Nikolic, I. and Sanchez, A. (2016). Institutional analysis for flood risk reduction: A coupled agent-based-flood models method. *8th International Congress on Environmental Modelling and Software (iEMSs2016)*, 10-14 July, Toulouse, France.

Abebe, Y.A., Vojinovic, Z., Nikolic, I., Hammond, M., Sanchez, A. and Pelling, M. (2015). Holistic flood risk assessment using agent-based modelling: the case of Sint Maarten Island. *EGU General Assembly 2015 (EGU2015)*, 12-17 April; Vienna, Austria.

Abebe Y.A., Seyoum S.D., Vojinovic Z., and Price R.K. (2014). Comparison of 2D numerical schemes for modelling supercritical and transcritical flows along urban floodplains. *11th International Conference on Hydroinformatics (HIC 2014)*, 17-21 August, New York City, USA

Abebe, Y.A., Vojinovic, Z., Seyoum, S.D. and Price, R.K. (2013). Comparison of numerical schemes for modelling supercritical flows along urban floodplains. *International Conference on Flood Resilience: Experiences in Asia and Europe*, 5-7 September, Exeter, UK.

Newsletter article

Abebe, Y.A., Medina Peña, N.J., and Vojinovic, Z. (2018). Strengthening Sint Maarten: Lessons Learned after Hurricane Irma. *Research Counts*, 2(12). Boulder, CO: Natural Hazards Center, University of Colorado Boulder. Available at: http://bit.ly/RC_SXM_Irma

Netherlands Research School for the
Socio-Economic and Natural Sciences of the Environment

DIPLOMA

for specialised PhD training

The Netherlands research school for the
Socio-Economic and Natural Sciences of the Environment
(SENSE) declares that

Yared Abayneh Abebe

born on 18 July 1985 in Addis Ababa, Ethiopia

has successfully fulfilled all requirements of the
educational PhD programme of SENSE.

Delft, 3 December 2020

Chair of the SENSE board

Prof. dr. Martin Wassen

The SENSE Director

Prof. Philipp Pattberg

The SENSE Research School has been accredited by the Royal Netherlands Academy of Arts and Sciences (KNAW)

KONINKLIJKE NEDERLANDSE
AKADEMIE VAN WETENSCHAPPEN

The SENSE Research School declares that Yared Abayneh Abebe has successfully fulfilled all requirements of the educational PhD programme of SENSE with a work load of 46.0 EC, including the following activities:

SENSE PhD Courses

- Environmental research in context (2015)
- Research in context activity: 'Co-organizing 36th IAHR World Congress on: Deltas of the future and what happens upstream (World Forum, The Hague – 28 Jun to 3 Jul 2015)'

Other PhD and Advanced MSc Courses

- Introduction to Complexity, Santa Fe Institute (2014)
- Complex Systems Summer School 2015, Santa Fe Institute (2015)

Societal impact activity

- Communication for broader audience: 'Strengthening Sint Maarten: lessons learned after hurricane Irma', https://hazards.colorado.edu/news/research-counts/strengthening-sint-maarten-lessons-learned-after-hurricane-irma (2018)

Management and Didactic Skills Training

- Supervised an MSc student with thesis entitled: "Institutional dimension of flood risk: Understanding institutional complexity in Flood Risk Management for the case of St Maarten" (2016-2017)
- Teaching in the MSc module 'Groupwork Sint Maarten' (2014-2015)
- Teaching in the MSc course 'Urban Flood Management and Disaster Risk Mitigation' (2019)
- Vice President of IAHR Young Professionals Network Delft (2014 -2016)
- Co-organised a half-day LaTex workshop for IHE Delft PhD researchers (2015)

Oral Presentations

- *Comparison of 2d numerical schemes for modelling supercritical and transcritical flows along urban floodplains*. 11th International Conference on Hydroinformatics, 17-21 August 2014, New York, United States of America
- *Institutional analysis for flood risk reduction: a coupled agent-based-flood models method*. 8th International Congress on Environmental Modelling and Software, 10-14 July 2016, Toulouse , France
- *Agent-based simulation for flood risk assessment*. 12th International Conference on Hydroinformatics, 21-26 August 2016, Incheon ,Korea

SENSE coordinator PhD education

Dr. ir. Peter Vermeulen